THE ESSENTIALS OF
VOLUMETRIC ANALYSIS

By the same authors

A. Holderness and J. Lambert

SCHOOL CERTIFICATE CHEMISTRY

A NEW CERTIFICATE CHEMISTRY

GRADED PROBLEMS IN CHEMISTRY TO ORDINARY LEVEL (with and without answers)

A CLASS BOOK OF PROBLEMS IN CHEMISTRY TO ADVANCED LEVEL (with and without answers)

PROBLEMS AND WORKED EXAMPLES IN CHEMISTRY TO ADVANCED LEVEL

WORKED EXAMPLES AND PROBLEMS IN ORDINARY LEVEL CHEMISTRY

THE ESSENTIALS OF QUALITATIVE ANALYSIS

A SIMPLE APPROACH TO ATOMIC THEORY

By A. Holderness

INTERMEDIATE ORGANIC CHEMISTRY

A SIMPLE APPROACH TO ORGANIC CHEMISTRY

INORGANIC AND PHYSICAL CHEMISTRY

REVISION NOTES IN ADVANCED LEVEL CHEMISTRY

    Book One: Organic
    Book Two: Inorganic
    Book Three: Physical

By J. Lambert

TABLES FOR ELEMENTARY ANALYSIS

By T. Muir and J. Lambert

PRACTICAL CHEMISTRY

By F. Sherwood Taylor

TABLES FOR QUALITATIVE ANALYSIS

ORGANIC CHEMISTRY

The Essentials of Volumetric Analysis

SECOND EDITION

by

JOHN LAMBERT
M Sc

in conjunction with
A. HOLDERNESS
M Sc, F R I C
and
F. SHERWOOD TAYLOR
M A, B Sc, Ph D

HEINEMANN EDUCATIONAL
BOOKS LTD · LONDON

Heinemann Educational Books Ltd
LONDON EDINBURGH MELBOURNE TORONTO
SINGAPORE JOHANNESBURG AUCKLAND IBADAN
HONG KONG NAIROBI NEW DELHI

ISBN 0 435 65534 5

© J. Lambert 1971

First published 1938
Reprinted 1941, 1942, 1943, 1944, 1945, 1946 (twice),
1948, 1949, 1951, 1953, 1954, 1955, 1956, 1957, 1958,
1959, 1960, 1961, 1963, 1965, 1966 (twice), 1968,
Second Edition 1971

Published by
Heinemann Educational Books Ltd
48 Charles Street, London W1X 8AH
Printed in Great Britain by
C. Tinling & Co. Ltd, Prescot

Contents

	Preface	vii
I.	Introduction	1
II.	Indicators	11
III.	Acidimetry and Alkalimetry	21
IV.	Potassium Permanganate	37
V.	Potassium Dichromate	49
VI.	Iodine and Sodium Thiosulphate	54
VII.	Silver Nitrate	61
VIII.	Potassium Thiocyanate	70
IX.	Adsorption Indicators	74
	Answers to Calculations	79
	Table of Atomic Weights	81
	Index	83
	Table of Logarithms	87

Preface to the Second Edition

In recent years most Boards have changed to the molar system of describing the concentrations of solutions though some have, as a temporary measure, not completely abandoned the normal system. It is thought that the molar system is sufficiently superior and widely accepted to justify switching all the instructions of the first edition of this book which has enjoyed many years of popularity. A few experiments thought too dangerous have been deleted.

The authors are grateful to Mr. J. S. Clarke of Alleyn's School, London, for his assistance in preparing this edition.

J. L.

Preface to the First Edition

THIS book is intended for use in the preparation of students for the G.C.E. Advanced Level examinations in Practical Chemistry. It is hoped that all the essential experiments are included, and that some help is given to scholarship candidates who often have to work on their own. Wherever it is thought that modern practice can be used to advantage in schools, the modern method is given. A short chapter on adsorption indicators is included.

The exercises given at the end of chapters contain many practical questions taken from scholarship papers and from the practical examinations of the various G.C.E. Boards.

There are many exercises of a more advanced character suitable for scholarship candidates or for students taking the course a second time. Some of these require no preparation on the part of the teacher, but can be worked completely by the student: these are marked with an asterisk so that they can be set in the knowledge that the student can carry on by himself.

We have pleasure in thanking Messrs. John Murray for permission to use experiments from the Science Masters' Book and the Examination Boards for permission to use their questions. The following is a list of these Boards and of the abbreviations used at the end of questions set by them to indicate the source of each particular question:—

University of London	(L.)
Joint Matriculation Board ...	(J.)
Oxford Local	(O.)
Oxford and Cambridge ...	(O. and C.)

We also thank the Examination Syndicates of the Universities of Oxford and Cambridge for permission to use some of their scholarship questions (Schol.) and Messrs. A. S. Langley and A. W. Wellings for proof-reading and for valuable suggestions.

J. L.

1. Introduction

As its name implies, volumetric analysis relies on methods involving accurate measurement of volumes of liquids, though one or more weighings may also be needed. Gravimetric analysis involves only weighings. Of the two methods of analysis, gravimetric analysis is the more accurate but volumetric analysis is much more rapidly carried out. Volumetric analysis is, however, by no means inaccurate and the error involved in an analysis carried out by an experienced worker should not exceed 0·2% but in a typical school experiment is about 1%.

Standard Solutions. Molarity

In general, a volumetric analysis is carried out by preparing a standard solution of the given material (or using a solution supplied) and determining the volume of it needed to react exactly with a known volume of another solution of accurately known concentration, in a chemical reaction for which the equation is known. The course of the reaction is traced by some means, usually by an indicator showing change of colour when the reaction is complete. To take a simple case, the concentration of a solution of potassium hydroxide could be determined by adding to it the indicator methyl orange, and then causing the alkali to react with a standard solution of hydrochloric acid slowly added until the solution just becomes orange, the acid being then just in excess. This process of adding one standard solution to another to determine equivalent volumes is called *titration*.

A *standard solution* is one of which the concentration is known. Any kind of unit of weight or volume may be used, but for scientific purposes grammes per litre (1 dm^3) is the most convenient.

For volumetric analysis, the system of working in *molar* solutions (or some multiple or submultiple of this concentration) is almost universal. A molar solution is a particular kind of standard solution and is defined in the following way:—

> A molar solution of a substance is one which contains one mole of the substance in 1 dm^3 (= 1000 cm^3) of solution.

For convenience such an expression as "a molar solution of sodium hydroxide" is usually written concisely as "M.NaOH." Double molar (2M), half molar (0·5M), decimolar (0·1M) and centimolar (0·01M) solutions are also commonly used and have the appropriate multiple or sub-multiple of the concentration of a molar solution.

A *litre* is the special name for a cubic decimetre of material and is 1000 cm^3; it should not be used where the results are of high precision.

A *mole* is the amount of substance of a system which contains as many elementary units as there are carbon atoms in 0·012 kilogramme of carbon-12. The elementary unit must be specified: it may be a molecule, an atom or an ion.

The *Avogadro Constant* is the number of particles in 0·012 kg of carbon-12, or one mole of any other material, and is 6·02 × 10^{23}.

(1) *Sulphuric Acid.*

H_2SO_4; molecular weight is 98, a molar solution contains 98 g/dm^3 (on the old system of normality there are 2 g replaceable hydrogen in 98 g of acid and so N.H_2SO_4 contains 49 g of the acid per dm^3).

(2) *Sodium Hydroxide.*

NaOH; molecular weight is 40, a molar solution contains 40 g/dm^3.

(3) *Potassium Permanganate.*

This compound, when reacting in acid solution (p. 37), gains electrons according to the basic equation:—

$$MnO_4^- + 8H^+ + 5e^- \rightarrow Mn^{2+} + 4H_2O$$

A molar solution would contain 158 g/dm^3 but such a solution cannot be used in practice because of limitations imposed by the low solubility of the compound; 0·02M KMnO$_4$ solutions, containing 3·16 g/dm^3 are usually employed.

(4) *Silver Nitrate.*

AgNO$_3$; molecular weight is 170, a molar solution would contain 170 g/dm^3 but usually 0·1M AgNO$_3$ (17 g/dm^3) is used.

The justification for the concentrations of other solutions will be found at appropriate places in the text. It should be noted particularly that the **concentration of a molar solution must be calculated from the formula of the compound allowing for any water of crystallization.** This is most important when a compound is capable of existing in two or more different forms. Consider the following varieties of sodium carbonate.

(1) Sodium carbonate-10-water (decahydrate). The molecular weight corresponding to $Na_2CO_3,10H_2O$ is $(2 \times 23) + 12 + (3 \times 16) + 10(2 + 16) = 286$. It is usually employed in 0·05M solution, which will contain $0·05 \times 286 = 14·3$ g/dm^3.

(2) Sodium carbonate (anhydrous). The molecular weight corresponding to Na_2CO_3 is 106. It is usually employed in 0·05M solution which will contain 5·3 g/dm^3.

Sodium carbonate possesses two molecular weights each

appropriate to the particular variety of crystal; consequently, it is important to know the formula of the crystals in order to make up a standard solution.

It follows from the definition of molar solutions that they do not contain chemically equivalent weights of the compounds per dm^3 of solution and, consequently, that equal volumes of different molar solutions are not chemically equivalent to one another, unless the equation involves one molecule of each of those substances.

In the reaction
$$NaOH + HCl \rightarrow NaCl + H_2O$$
there is one molecule of each of the reagents but in the reaction
$$2NaOH + H_2SO_4 \rightarrow Na_2SO_4 + 2H_2O$$
the ratio of the number of molecules is 2 : 1. In 1 dm^3 (1000 cm^3) of a molar solution of a material there is 1 mole of the material thus in v cm^3 of an m.M solution there are $\dfrac{vm}{1000}$ moles of material. The indicator shows when two materials have reacted and at this point
$$\frac{v_A m_A}{v_B m_B} = \frac{a}{b}$$
if the equation is
$$aA + bB \rightarrow cC + dD$$
because the experimentally determined ratio of the number of moles of material is equal to the ratio of the numbers of molecules of the materials.

Instruments of Volumetric Analysis

The course of a volumetric analysis is usually the following:—
A standard solution of the material to be analysed is made up by weighing in *a weighing bottle* an appropriate quantity of the material, making it up to 250 cm^3 of aqueous solution in a *measuring flask*, transferring 25 cm^3 of this solution to a conical flask by means of a *pipette* and titrating it with a standard (often decimolar) solution of some reagent from a *burette*, using a suitable indicator. Each of these measuring instruments and the manner of its use will now be considered. A measuring cylinder may occasionally be used but only for very approximate work.

Weighing Bottle. This consists of a cylindrical glass vessel with an accurately ground stopper (Fig. 1), in which the materials can be weighed out of contact with the open atmosphere. The weighing bottle is usually heated in a steam oven before use to ensure dryness and is then allowed to cool in a desiccator. It should be handled in a dry cloth to avoid contamination with grease from the fingers. It may be used in one of two ways. The first method is to weigh the bottle empty, powder the material given in a clean dry mortar and then weigh out an amount of material exactly. This has the disadvantage

4 THE ESSENTIALS OF VOLUMETRIC ANALYSIS

of being tedious and exposing the material to the atmosphere while adjustments of amount are being made. A spatula of plastic material or stainless metal is used to transfer the substance. The second method is to weigh the bottle containing an amount of the material

Fig. 1.

known to be roughly suitable for the purpose in hand, to transfer the material to a measuring flask as described in the next paragraph and weigh the bottle containing a trace of residual material, after which the actual weight transferred to the measuring flask can be obtained by difference. The second method is much quicker and is preferable except where solutions of an exact molarity are being prepared directly: the first method is then essential. Plastic weighing bottles can frequently be substituted for the more fragile glass ones.

The degree of accuracy necessary in the weighings will be considered later (p. 9).

Measuring Flask. Measuring flasks of 250 cm^3 capacity are usually employed (Fig. 2) because the amount of solution used in a single titration is usually 25 cm^3 and several such titrations may be carried out by drawing from the 250 cm^3 of solution prepared. Measuring flasks of 100 cm^3, 500 cm^3, 1000 cm^3 and 2000 cm^3 are also in frequent use.

It should be noted that a measuring flask is made to *contain* a volume of liquid and will not *deliver* that volume, because some liquid is inevitably retained as a film on the sides of the flask. Measuring flasks are usually graduated at 20°C and should only be used at temperatures close to this.

When a solution is to be made up the measuring flask should first be well rinsed with several small quantities of the solvent (usually distilled water) that is to be used. This removes any traces of impurities. A small beaker should be similarly rinsed and the material transferred to it carefully from the weighing bottle. The solvent is then added from a wash-bottle down the sides of the

INTRODUCTION

beaker so that there is no splashing. Gentle stirring with a glass rod will hasten the process of solution, but the rod should not be removed from the beaker unless all solution is first washed from its surface into the beaker. (If the solvent is used hot, the solution must finally be cooled to room temperature. The solution must also, of course, remain unchanged by heat.) When the solute is completely dissolved the glass rod should be placed in a funnel which rests in the neck of the measuring flask and the solution poured down it from the beaker into the flask. The entire interior surface of the beaker should then be washed several times with the solvent and the washings transferred to the flask down the rod which will also be

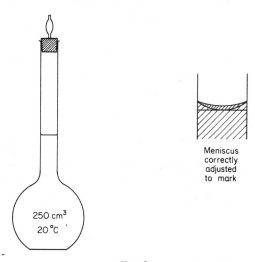

Fig. 2.

washed by them. (If the material is being weighed by difference, the weighing bottle should now be weighed again. If not, the weighing bottle should be washed out with solvent in the same way as the beaker and the washings added to the contents of the flask.) The measuring flask is then filled with the solvent from a wash-bottle until the bottom level of the meniscus is at the mark (Fig. 2). A pipette should be used to add the last drop or two. (When deciding whether the level is correctly adjusted, lower the eye until the mark is at your eye-level and so avoid error due to parallax.) The flask should then be stoppered and **shaken vigorously** for some few minutes to make the solution uniform. After shaking, the solution will be *below* the mark: some of the solution is retained as a film on the stopper and neck of the flask. The total volume is still, however, 250 cm³ and 25 cm³ taken from it will be accurately one-tenth. Do not be tempted to make the solution up to the mark again. This proceeding makes the volume 250 cm³ plus an unknown added volume.

If time is an important factor the solution can be made uniform more quickly by pouring it into a large dry flask in which it can easily be swirled round for a short time. **It is essential that the solution should be of uniform concentration throughout.** Whenever the words "shake well" appear in the text they mean this essential process of making the solution homogeneous.

Pipette. The pipette (Fig. 3) is designed to *deliver* a certain volume of liquid. When filled to the mark it contains more than this volume, a little of the liquid being retained after delivery as a film on the sides of the pipette and in the tip. The actual volume *delivered* from the pipette should be constant, and it is therefore important to observe certain conditions when using the pipette so that the small volume of liquid retained in it is constant.

The pipette is filled above the mark by sucking solution into it and this liquid is allowed to drain away. A pipette with suction device or a burette should always be used for poisonous or corrosive solutions. If a liquid is taken into the mouth it should be spat out at once and the mouth washed out. This process of filling and allowing the liquid to run away should be repeated to ensure that the pipette contains nothing but the solution which is to be measured. The pipette is then filled above the mark and the liquid is retained by pressing the forefinger on the open end of the stem. The pipette is then raised so that the mark is at eye-level and by controlled release of the finger from the stem, the liquid is allowed to fall slowly until the bottom level of the meniscus is at the mark. The tip of the pipette is then placed inside a conical flask which should be dry or should contain only distilled water. By removing the finger from the stem, the liquid may be delivered from the pipette into the flask. A 25 cm³ pipette should deliver its contents in about 20 seconds and the point of the pipette should be touching the side of the vessel. If delivery is more rapid than this, the volume delivered is not constant. A pipette which delivers too rapidly should have its extreme tip heated very gently in a flame, when the hole will close slightly. A little liquid will be retained in the tip; no attempt should be made to expel it as the pipette will already have *delivered* 25 cm³ of liquid which is now available for titration with a standard solution from a burette, using a suitable indicator. The greatest difference between deliveries from a 25 cm³ pipette should not exceed 0·025 cm³.

Burette. The burette is illustrated in Fig. 4. Burettes employed in volumetric analysis usually have a capacity of 50 cm³ and are graduated in cm³ and 0·1 cm³. A glass-stoppered burette is to be preferred and the stopper should turn smoothly. If it sticks, the socket and stopper should be dried and a *very thin* smear of vaseline

INTRODUCTION

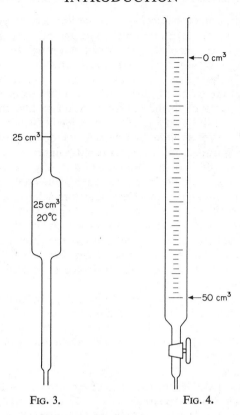

Fig. 3.   Fig. 4.

placed on the stopper. If too much is used, the vaseline may block the hole in the stopper. A rubber band may be used to prevent the stopper from sliding out of the socket. A Mohr burette is quite satisfactory for most purposes, but it should not be used with iodine solutions. The rubber connection is attacked by iodine. The burette is first washed out with the solution it is to contain, the washings being allowed to run away *through the jet* so as to wash this part also. A second washing is desirable to ensure complete elimination of impurities. It is then filled up to the region of the zero mark with the solution and *the jet is filled* by opening the tap for a second or two. Time is then allowed for drainage down the sides of the burette, after which with the surface of the liquid at eye-level, the reading of the *bottom level* of the meniscus is taken. A white sheet of paper held behind the liquid at an angle of 45 degrees will help to define the meniscus and a good approximation to the second decimal place may be obtained. The titration is then completed and the new reading taken, after which the volume of liquid delivered is found by difference.

8 THE ESSENTIALS OF VOLUMETRIC ANALYSIS

As a titration must be accurate to one drop of reagent and the volume of it needed is at first only very approximately known, it is almost always a saving of time to carry out a rough titration first in the following way, after which accurate titration can be quickly performed. To the solution and indicator in the conical flask, add the solution from the burette 1 cm^3 *at a time*, until excess is present as shown by the change in the indicator. Suppose that the indicator changes between the 23rd and 24th cm^3 added. Then in subsequent titrations 22 cm^3 of solution can be safely added from the burette, after which adjustment to the accurate end-point must be made drop by drop: the smallest possible addition at this slow rate will be about 0·05 cm^3. It is usually quicker to carry out a deliberately approximate titration first, although with some indicators a fairly accurate first titration is possible because the indicator shows signs of the arrival of the end-point. In the second and subsequent titrations, the flask should be shaken at intervals and the reagent should be added in quantities of not more than about 5 cm^3 because the large local excess of reagent which may result is apt to induce undesirable variants of the main reaction.

A burette (including the tap) should always be well washed out after use. If an alkaline solution has been in the burette, about 10 cm^3 of dilute acid should be run into the burette after running out the alkali, and the burette then well washed out with water.

Sources of Error

The following are the chief sources of error in volumetric analysis.

Solution not Homogeneous. This is a frequent source of error. After the solution has been made up to the mark, it is essential either to shake very well or to pour the solution into a large flask and swirl.

Inaccuracy of Instruments Used. Measuring flasks, burettes and pipettes of reasonable price are necessarily manufactured by mass-production methods and inaccuracies are certain to arise. An experienced analyst can calibrate his apparatus and so practically eliminate errors from this source. Actually the least accurate of the instruments is the burette for the following reasons:—

(1) An error of a drop may arise in the titration because this is the least amount that can be added. The volume of a drop from an ordinary school burette may be about 0·05 cm^3, but the error may be reduced by averaging three close titrations.

(2) The burette may drain irregularly. For this reason, burettes should be treated at intervals with a cleaning mixture to remove grease.

(3) The second decimal place in the readings can only be obtained approximately.

The probable error in using the burette is about 1 part in 500, and this is usually the greatest error an experienced analyst will encounter. The other instruments are generally more accurate, provided that they are consistently used as previously described in the text.

Errors in Weighing. As the error in using the burette is about 1 part in 500, there is no point in weighing out material extremely accurately. If one gramme or more is being weighed, a mistake of one unit in the third decimal place will introduce an error of one part in 1000, which is much less than the burette error. Thus a weighing of one gramme or more of material should not be carried beyond the third decimal place.

Impurity of Materials. It is obvious that if analyses are based on solutions made up from impure chemicals, the results will be unreliable. There has, however, been a remarkable improvement in recent years in the quality of analytical reagents, and chemicals of analytical reagent quality are now so pure that the errors they introduce, compared with those from other sources, are almost always negligible. Moisture is nearly always present to the extent of 0·2 to 0·5% in any powder not specially dried. This impurity can usually be removed by storing the substance before use for a few hours in a desiccator.

Errors may arise from the action of light, atmospheric carbon dioxide, dust particles or oxygen on standard solutions. These may be minimised by the use of coloured glass bottles, tightly fitting stoppers, soda-lime tubes to absorb carbon dioxide and in other ways. If a solution is kept for a long time, it should be standardised at intervals.

Inaccuracy in the End-point Recorded

If too much indicator is added to the solution to be titrated, a certain amount of the reagent added will be needed to cause the colour change to take place. It is often helpful to perform a "blank" experiment to ascertain the volume of reagent necessary to affect the indicator which has been added to a volume of water approximately equal to the final volume of solution likely to be obtained in the actual titration. Furthermore, where a change in colour indicates the end-point of a reaction, practice is often necessary before the change can be clearly recognised. It is sometimes useful to have on the bench for comparison a flask containing a few drops of the unchanged indicator added to water, so that any alteration in colour can be easily observed.

Accuracy

The results of one person performing the same experiment many times or of many people doing the same experiment once will fall in a pattern known as a normal distribution curve. The sharper the peak the more accurately the experiment has been performed (Fig. 5).

10 THE ESSENTIALS OF VOLUMETRIC ANALYSIS

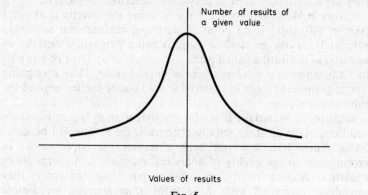

Fig. 5.

The results obtained by volumetric analysis are usually accurate to 3 significant figures but if the sequence of digits is above 500 then it is more realistic to quote only 2 significant figures because of the unreliability of the third one.

2. Indicators

Ionisation

THE aqueous solutions of some substances will readily conduct an electric current and decomposition occurs as a result. These substances are termed electrolytes, and acids, alkalis, and the majority of salts belong to this class. Thus a solution of hydrochloric acid in water conducts an electric current and is decomposed into hydrogen (which is evolved at the negative pole or cathode) and chlorine (at the anode or positive pole). This is explained by assuming that ions are present in the liquid before any current is passed through it. These ions are electrically charged atoms or groups of atoms, ions of metals or metallic groups being positively charged and ions of non-metals and non-metallic groups being negatively charged. The amount of charge is directly proportional to the valency of the atom or group. Since the majority of the volumetric processes depend upon the interaction of ions we shall often represent a substance, not by its usual formula, but by its ionic formula. This indicates the ions furnished by that substance when dissolved in water. Thus:—

HCl ⎫   ⎧ $H^+ + Cl^-$
H_2SO_4 ⎪   ⎪ $2H^+ + SO_4^{2-}$
$NaOH$ ⎬ may be represented ⎨ $Na^+ + OH^-$
$Ca(OH)_2$ ⎪ as indicated opposite. ⎪ $Ca^{2+} + 2OH^-$
KNO_3 ⎪   ⎪ $K^+ + NO_3^-$
$BaCl_2$ ⎭   ⎩ $Ba^{2+} + 2Cl^-$

The plus sign between them merely indicates that the ions are present in the same solution in those proportions. It does not imply any chemical bond between the ions. They are free to move anywhere in the solution. When an electric current is passed, however, the negative ions or anions move to the positive pole and the positive ions or cations to the negative pole; and the phenomenon of electrolysis is observed.

Definitions. *An acid* is a substance which contains hydrogen and when dissolved in water furnishes hydrogen ions,

e.g., $HCl = H^+ + Cl^-$

A base is a substance which will react with hydrogen ions to give a salt and water only. The alkalis are substances which when dissolved in water furnish hydroxyl ions,

e.g., $NaOH = Na^+ + OH^-$

12 THE ESSENTIALS OF VOLUMETRIC ANALYSIS

Neutralisation is a reaction between an acid and a base producing a salt and water only, *e.g.*,

$$Na^+ + OH^- + H^+ + Cl^- = Na^+ + Cl^- + H_2O$$

It will be clearly seen that, since the common salt produced is in the form of ions of sodium and chlorine, neutralisation consists essentially of the action between hydrogen ions and hydroxyl ions to form molecules of water which are undissociated. (See p. 13.)

$$H^+ + OH^- = H_2O$$

Weak and Strong Electrolytes. According to Arrhenius the ions are furnished by a reversible reaction in which an undissociated molecule splits up into ions to an extent which varies considerably from substance to substance and according to the dilution of the solution. The greater the dilution the greater the dissociation.

Strong electrolytes are dissociated to a considerable extent even in fairly concentrated solution, whereas weak electrolytes are only slightly dissociated. Strong electrolytes are believed to be completely dissociated even in concentrated solution, and it is because their mobility (*i.e.*, the free movement of the ions) is restricted in concentrated solution that they appear to be to some extent undissociated. The following table shows how the dissociation varies for two weak electrolytes and also how the dilution affects the degree of dissociation.

TEMPERATURE 18°C DEGREE OF DISSOCIATION

	0·1M	0·01M	0·001M
Ammonia	0·0133	0·0415	0·125
Acetic acid	0·0133	0·0415	0·125

Thus in 0·1M acetic acid 1·33% of the weight of the acetic acid consists of hydrogen ions and acetate ions.

Dissociation Constant. Consider an electrolyte AB which ionises when in solution.

$$AB \rightleftharpoons A^+ + B^-$$

The equilibrium constant, K, is given by

$$K = \frac{(\text{Conc. of A}^+)(\text{Conc. of B}^-)}{(\text{Conc. of unionised AB})}$$

where the concentrations are expressed in mole/dm³. The equation is often expressed thus:—

$$K = \frac{[A^+][B^-]}{[AB]}$$

The equilibrium constant in this case is termed the dissociation constant of the electrolyte. This can be calculated from the above table thus:—

Dissociation Constant for Acetic Acid.

$$CH_3CO_2H \rightleftharpoons H^+ + CH_3CO_2^-$$

Conc. in 0·1M sol.:   0·09867   0·00133   0·00133
in mole/dm³

$$\therefore K = \frac{(0 \cdot 00133)^2}{0 \cdot 09867} = 1 \cdot 8 \times 10^{-5}$$

Ionic Product of Water. We have seen that neutralisation consists of the formation of molecules of undissociated water by the chemical union of hydrogen ions and hydroxyl ions.

$$H^+ + OH^- = H_2O$$

Like other ionic reactions this is to some extent reversible.

It is true that pure water is practically a non-conductor of electricity, but from conductivity experiments it can be shown that pure water does contain minute quantities of both hydrogen ions and hydroxyl ions. Then:

$$H_2O \rightleftharpoons H^+ + OH^-$$

represents the equilibrium between the ions and undissociated water, and the dissociation constant K would be given by

$$K = \frac{[H^+][OH^-]}{[H_2O]}$$

The amount of the ions is so small, however, that $[H_2O]$ can be considered to be constant without serious error and the expression $[H^+][OH^-]$ is therefore also constant and is termed the ionic product of water. The value of this product at 25°C is very nearly 10^{-14}, and since the ions are present in equal amounts (1 molecule of water gives 1 molecule of hydrogen and 1 molecule of hydroxyl ion) it follows that pure water contains a concentration of hydrogen ion of 10^{-7} g/dm³.

$$[H^+][OH^-] = 10^{-14} \text{ (at 25°C)}$$

The concentrations are expressed thus:—

$[H^+] = 10^{-7}$ mole/dm³ or 10^{-7} g of hydrogen ions per 1000 cm³.

$[OH^-] = 10^{-7}$ mole/dm³ or 17×10^{-7} g of hydroxyl ions per 1000 cm³.

Although this ionic product is exceptionally small it is of great importance because it is this constant which is used to trace the alteration of the hydrogen ion concentration which takes place during neutralisation.

It is important to note that however strongly acidic an aqueous solution may be, it is never completely free from OH^- because there must be enough OH^- to satisfy the equation:—

$$[H^+][OH^-] = 10^{-14}$$

Similarly, an alkaline solution is never completely free from H^+. When H^+ and OH^- are present in equal proportions, the concentration of each can only be 10^{-7} mole/dm³, and this constitutes a neutral solution.

Hydrolysis. The presence in water of hydroxyl ions and hydrogen ions accounts for the phenomenon of hydrolysis. This action can be considered as the reverse of neutralisation and occurs when the ions of water react with the ions derived from a substance in solution. Thus, sodium carbonate in solution furnishes carbonate ions and sodium ions.

$$Na_2CO_3 \rightleftharpoons 2Na^+ + CO_3^{2-}$$

We have present in sodium carbonate solution the following ions:

$$Na^+, CO_3^{2-}, H^+, OH^-$$

although the last two are only present to a very small extent. Now carbonic acid is a very weak acid and dissociates only very slightly, hence the following reaction occurs until there is left only that small quantity of hydrogen ions necessary to maintain the very small equilibrium constant for carbonic acid:

$$2H^+ + CO_3^{2-} \rightleftharpoons H_2CO_3$$

As hydrogen ions are removed from the sphere of action by the formation of this undissociated acid more hydrogen ions are liberated by the dissociation of water to maintain the value of the ionic product

$$[H^+][OH^-] = 10^{-14}$$

Finally an equilibrium is set up leaving in solution an excess of hydroxyl ions over hydrogen ions which causes the solution to be alkaline.

By a similar type of reasoning it can be shown that a sodium hydrogen carbonate solution (which is a solution of an *acid* salt) is alkaline. Thus:—

$$H_2O \rightleftharpoons H^+ + OH^-$$
$$HCO_3^- + H^+ \rightleftharpoons H_2CO_3$$
Only slightly dissociated.

The removal of the hydrogen ions to form undissociated carbonic acid leaves an excess of OH^-, hence the solution is alkaline.

Similarly ammonium chloride furnishes ammonium and chloride ions in solution.

$$NH_4Cl = NH_4^+ + Cl^-$$

Ammonium hydroxide is only slightly dissociated and hence the ammonium ion reacts with the hydroxyl ions present to form ammonia and water.

$$NH_4^+ + OH^- = NH_3 + H_2O$$

This removes hydroxyl ions from the solution, leaving an excess of hydrogen ions which cause the solution to be acidic.

It follows that the solution of a normal salt in water, *i.e.*, a salt formed by replacing all the replaceable hydrogen by a metal, is by no means certain to be neutral. Whether such a salt will form an acidic or an alkaline solution can usually be predicted by an examination of the possible reactions of its ions with hydrogen ions or hydroxyl ions.

pH Value. It was suggested by Sørensen that a useful method of denoting the hydrogen ion concentration of a solution would be by

INDICATORS

using the logarithm of the concentration with the sign reversed.
By this means the concentration of an acidic or an alkaline solution can be expressed in the same terms, since if the pH value is known the concentration of the hydroxyl ion can be at once determined from the equation:—

$$[H^+][OH^-] = 10^{-14}$$

By definition

$$pH = -\log[H^+]$$

e.g., let $[H^+] = 10^{-2}$ mole/dm^3, *i.e.*, 0·01 g/dm^3 as would be the case in a solution of a strong acid of concentration 0·01M. It is assumed that the strong acid is completely dissociated.

Then
$$\log[H^+] = -2$$
$$\therefore pH \text{ value} = 2$$

pH *Value of* 0·1M *Acetic Acid*

In 0·1M acetic acid, degree of dissociation = 0·0133

$$\therefore [H^+] = \frac{1}{10} \times 0{\cdot}0133 \text{ mole/dm}^3$$

$$= 0{\cdot}00133 \text{ mole/dm}^3$$

pH value $= -\log 0{\cdot}00133$
$= -[\overline{3}{\cdot}1239]$
$= --[2{\cdot}8761]$
$= 2{\cdot}87$

(Compare 0·001M HCl; pH value = 3·00, assuming complete dissociation.)

pH *Value of* 0·01M *Sodium Hydroxide*.

$$[OH^-] = \frac{1}{100} \text{ or } 10^{-2} \text{ mole/dm}^3$$

but
$$[OH^-][H^+] = 10^{-14} \text{ at } 25°C$$
$$\therefore [10^{-2}][H^+] = 10^{-14}$$
$$\therefore [H^+] = 10^{-12}$$
$$\therefore pH = 12$$

It follows that an exactly neutral solution is one of pH value 7, since this solution will contain equal amounts (expressed as moles) of hydrogen ion and hydroxyl ion. The smaller the pH value the more acidic the solution and the larger the pH value the more alkaline the solution.

Another advantage of using the pH system is that changes in concentration too large to be represented graphically can easily be traced by using pH value. As an example, consider the addition of 25 cm^3 of M HCl to 25 cm^3 M NaOH. The hydrogen ion concentration of the alkaline solution if completely ionised is 10^{-14} mole/dm^3, whereas that of the M acid is 1 mole/dm^3 and at exact neutrality the hydrogen ion concentration is 10^{-7} mole/dm^3.

16 THE ESSENTIALS OF VOLUMETRIC ANALYSIS

Using pH value this change can be shown graphically as in Fig. 6. It can be seen there is a rapid change of pH value as the solution changes from being slightly alkaline to being slightly acidic.

To calculate change of pH value as one drop of excess acid falls into the neutral solution.

If 25 cm^3 M HCl have been added to 25 cm^3 M.NaOH the solution will be exactly neutral and its pH value will be 7. If 1 drop = 0·05 cm^3 it will contain 0·05 × 10^{-3} g of hydrogen ions. This is added to 50 cm^3 of neutral solution, and hence

$$[H^+] = \frac{1000}{50} \times 0.05 \times 10^{-3} \text{ g/dm}^3$$

$$= 10^{-3} \text{ g/dm}^3$$

∴ pH value = 3

Neglecting the readjustment due to the existing small amount which is merely 10^{-7} g/dm^3 we see that one drop of acid *at this stage* produces an increase in [H$^+$] of 10 000 times its value. By a similar argument it can be shown that a further 10 drops (0·5 cm^3) would be required to change the pH value to 2.

Theory of Indicators. Most of the indicators used in acidimetry and alkalimetry are weak acids and their degree of dissociation is greatly affected by alteration of hydrogen ion concentration of the solution, thus producing a change in colour. Consider phenolphthalein, a very weak organic acid HA, its dissociation can be regarded as the splitting up of the molecule into a hydrogen ion and an ion negatively charged.

$$HA \rightleftharpoons H^+ + A^-$$
(colourless)   (pink)

The colour change of an indicator may not be due solely to changes in extent of ionisation but this is a good approximation. If dissociation occurs to any marked extent it is clear that a pink colour will be observed whereas if the undissociated acid is mainly present no colour will be observed. Consider the effect of adding an acidic solution, the hydrogen ion concentration of which is high (compared with the concentration of H$^+$ furnished by the indicator) so the dissociation of the indicator is suppressed. Hence the indicator will be practically unionised and no colour will be observed. On adding an alkaline solution, however, hydrogen ions from the indicator will react with some of the hydroxyl ions present to form undissociated water, and hence the dissociation of the weak acid will be increased with a corresponding increase in the coloured ions A$^-$ which will make the solution pink. Colour changes such as this take place at different concentrations with different indicators, and below is a list of some of the indicators in common use together with the range of pH value over which the colour change takes place.

INDICATORS

pH range	Indicator	Colour change Acid-alkali
3·0–4·4	Methyl orange	Red-yellow
4·4–6·3	Methyl red	Red-yellow
6·0–7·6	Bromo-thymol blue	Yellow-blue
6·0–8·0	Litmus	Red-blue
8·2–10·0	Phenolphthalein	Colourless-red

It is quite obvious from Fig. 6 that any of these indicators will give a sharp end point if a strong base is being titrated with a strong acid, because the addition of only 1 or 2 drops of acid causes the pH value to move over the whole of the vertical portion of the curve.

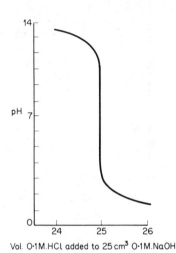

Vol. 0·1M.HCl added to 25 cm³ 0·1M.NaOH

FIG. 6.

Essential Characteristics of a Good Indicator.

(1) The colour change of the indicator must be clear and sharp, *i.e.*, it must be sensitive. Thus it would be useless if 2 or 3 cm³ of the reagent were necessary to bring about the colour change.

(2) The pH range over which the colour change takes place must be such as to indicate when the reaction (as shown by the equation) is complete.

Choice of Indicator

(*a*) **For Titration of Weak Acids with Strong Alkalis.** We have already seen, p. 12, that a weak acid is only slightly dissociated, and hence its hydrogen ion concentration is low. Now a strong acid in 0·1M solution will have a pH value of about 1, and we will suppose that our weak acid has in 0·1M solution a pH value of 4. It would still be definitely on the acid side of the neutral point. Now if we titrate this weak acid with 0·1M alkali (pH value 13) it is clear that no matter how much acid we added we could not, for example, obtain a solution of pH value 3.

In other words the range of pH value of the solutions as the two are mixed cannot be outside the limits pH = 4 to pH = 13. For this reaction it would be useless to choose an indicator which changed at pH value 3. It would show no change. We can go further. We should not think it advisable to choose an indicator which changed

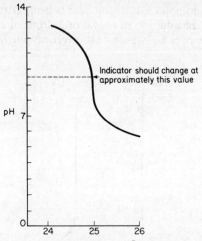

Vol. 0·1M.CH$_3$CO$_2$H added to 25cm^3 0·1M.NaOH

FIG. 7.

at pH value 5 because the pure acid itself would have to be present in a concentration somewhere near 0·01M to produce this effect. The best indicator to use would be one which changed at about pH value 9. The graph, Fig. 7, shows the pH change as a weak acid is added to a strong alkali. For this titration the best indicator is phenolphthalein.

Hence, when titrating:—

Oxalic
Acetic
Succinic
Formic
Boric
Carbonic
Sulphurous
 or any weak acid
} against a strong alkali use phenolphthalein as an indicator.

If the acid is being run into the alkaline solution the disappearance of the pink colouration gives the correct end-point. Carbon dioxide of the air may dissolve sufficiently to cause the pink colour of the

INDICATORS

phenolphthalein to disappear if the flask is shaken too vigorously.

(*b*) **For Titration of Weak Alkalis with Strong Acids.** The case is reversed when we are considering the indicator to use if a strong acid and a weak alkali are to be titrated. The pH value of the weak alkali in the concentration used may not exceed 10 (*i.e.*, the solution could not have a high concentration of hydroxyl ions). The pH value cannot vary outside the limits 1–10 (for, say, decimolar solutions) and comparatively large concentrations of the alkali might be necessary to obtain a pH value anywhere approaching 10. Fig. 8

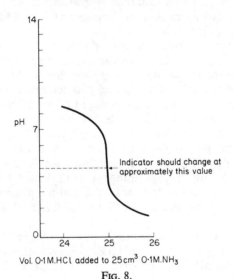

Fig. 8.

shows the pH change during a titration of a strong acid against a weak alkali. Hence for this type of reaction (weak alkali—strong acid) an indicator changing at a pH value of about 4 is essential. That used in practice is methyl orange or methyl red.

Hence when titrating Ammonia solution, Sodium carbonate, Sodium borate or any weak alkali against a strong acid use methyl orange or methyl red.

It is impossible to follow the neutralisation of a weak alkali by a weak acid as there is no point in the titration at which a rapid change of pH value takes place.

Notes on Some of the Indicators in General Use

Litmus contains several dyes and the colour change is therefore not easy to follow if very accurate work is being done. Furthermore, the solution does not keep well out of contact with air. Litmus contains azolitmin, and the latter is a very sensitive dye, and can with advantage be substituted for litmus.

Litmus Solution. [Colour change red-blue pH 6·0–8·0.]

Digest 10 g of commercial litmus with about 500 cm^3 of warm water and allow to stand for some time. Filter off and add to the filtrate a few drops of dilute nitric acid until the purple colour is obtained. Keep in a bottle with a loose cork.

Azolitmin. [Colour change red-blue pH 5·0–8·0.]

Dissolve about 5 g of azolitmin in 500 cm^3 of water to which a little sodium hydroxide solution has been added. Add dilute nitric acid until the purple colour is obtained.

Methyl Orange. [Colour change red-orange pH 3·0 to 4·4.]

This is an excellent indicator for use as described above if the concentration of the solutions is greater than 0·2M. In decimolar solutions the end-point is not too easy to determine.

Methyl Orange. 1 g of methyl orange to 500 cm^3 water. Use 2 or 3 drops for 25 cm^3 of solution to be titrated.

Methyl Orange modified with Xylene Cyanol F.F. [Colour change green—neutral grey—magenta.] Hickman and Linstead advocate mixing this dye with xylene cyanol F.F., which has the effect of producing with the methyl orange a neutral grey tint at the pH value 3·8 which corresponds with the end-point given by methyl orange. The colour change (with the addition of xylene cyanol F.F.) is:—

    Green   —   neutral grey   —   magenta
 alkaline side   pH 3·8   acid side of
  of pH 3·8   pH 3·8

Mix 1 g methyl orange with 1·4 g cyanol and 500 cm^3 of 50% alcohol and water. Use 2 or 3 drops for 25 cm^3 of solution to be titrated.

Methyl Red [colour change red-yellow pH 4·4 to 6·3] is a sensitive indicator and very useful for titrating weak organic bases and ammonia. Dissolve 1 g of the dye in 500 cm^3 of 60% alcohol. Use one or two drops only for 25 cm^3 of liquid to be titrated.

Phenolphthalein. [Colour change colourless-red pH 8·2 to 10·0.] For use see p. 18.

Dissolve 1 g in 500 cm^3 of 50% alcohol.

3. Acidimetry and Alkalimetry

BEFORE estimations involving acids or alkalis can be carried out on given substances or mixtures, it is necessary to obtain accurately standard acidic and alkaline solutions. The common acids and alkalis cannot be employed directly for making standard solutions because they are variable in composition for the reasons given.

Hydrochloric Acid. This is volatile in high concentrations.
Sulphuric Acid. This is hygroscopic.
Nitric Acid. This is volatile and subject to decomposition.

Furthermore, the so-called "pure" mineral acids of the laboratory are not pure substances: they contain varying quantities of water.

Sodium and Potassium Hydroxides. These are deliquescent and react with carbon dioxide from the air.

$$e.g., 2NaOH + CO_2 = Na_2CO_3 + H_2O$$

Calcium Hydroxide. This is insufficiently soluble and also reacts with carbon dioxide from the air.

$$Ca(OH)_2 + CO_2 = CaCO_3 + H_2O$$

Ammonia. This is volatile, and is a solution of variable concentration.

Characteristics of a good standardising reagent:—
(a) It should be obtainable in a high degree of purity.
(b) It should be stable and unaffected by the atmosphere. It should not be deliquescent or efflorescent, so that it may be weighed easily.
(c) It should be fairly cheap.

For standardisation of acids the materials commonly used are
(1) Pure sodium carbonate prepared by heating sodium hydrogen carbonate (or analytical quality anhydrous sodium carbonate).
(2) Pure borax $Na_2B_4O_7,10H_2O$, sodium metaborate.
(3) Iceland spar, pure calcium carbonate.

Alkaline solutions may be standardised by using solid crystalline organic acids such as oxalic acid or succinic acid, which can be obtained in a high state of purity.

Experiment 1. To standardise a solution of hydrochloric acid.
(Concentrated hydrochloric acid. Sodium hydrogen carbonate.)
Concentrated hydrochloric acid is roughly 11M. Pour out into a

22 THE ESSENTIALS OF VOLUMETRIC ANALYSIS

measuring cylinder about 2 cm³ of concentrated hydrochloric acid transfer to a 250 cm³ flask and make up to the mark with water. Shake well. Put some pure sodium hydrogen carbonate or anhydrous sodium carbonate into an evaporating dish and heat gently over a low flame for about 15 minutes. Take care not to heat the mass too strongly or fusion may take place which will seriously retard solution and also cause slight decomposition and stir continuously. It will be obvious when carbon dioxide is being evolved as the mass appears unusually light as it is being stirred. Allow the dish to cool in a desiccator because anhydrous sodium carbonate absorbs moisture to form the monohydrate and weigh. Heat again for five minutes, cool and weigh. Repeat this process until the weight is constant.

$$Na_2CO_3 + 2HCl = 2NaCl + H_2O + CO_2$$

Molecular weight 106.

Weight of Na_2CO_3 for 250 cm³ 0·05M solution = 0·25 × 0·05 × 106 = 1·325 g.

Weigh a clean dry weighing bottle and weigh out exactly 1·325 g of the pure sodium carbonate into it. Transfer the carbonate to a beaker containing a little water (shaking gently as the carbonate comes in contact with the water) and wash the weighing bottle carefully by means of the wash-bottle, allowing the washings to drop into the beaker. Stir gently to dissolve, warming if necessary. Then cool to room temperature before proceeding. Smear an almost imperceptible amount of vaseline on the lip of the beaker and pour the solution down a glass rod into a funnel resting in the neck of a clean (but not necessarily dry) 250 cm³ flask. Wash the beaker out with further small quantities of water, pouring all washings down the rod and funnel into the 250 cm³ flask to ensure that no solution is left on the walls of the beaker. It is well to remember that once you have weighed out the sodium carbonate into the weighing bottle every particle of it, whether as solid or as a solution, must be transferred into the 250 cm³ flask. Make up to the mark with water from a pipette, shake well or pour into a large dry beaker in order that the solution may become homogeneous.

Take 25 cm³ of the 0·05M sodium carbonate solution by means of a pipette, run it into a conical flask, add 2 or 3 drops of methyl orange solution, make a note of the burette reading and run in the acid from a burette. The first 15 to 18 cm³ of the acid may be run in without fear of overshooting the end-point. Then run in 1 cm³ at a time, shaking after each addition until the colour changes from yellow to pink. This is a trial titration and unless you are convinced of its accuracy it should not be used in the calculation. The titration should now be repeated with further portions of 25 cm³ of the carbonate solution until two readings are obtained which agree to 0·1 cm³ or three readings which do not show any trend in value, up or down.

ACIDIMETRY AND ALKALIMETRY

Readings of burette in cm³

Trial	1	2	3
Final	23·4	45·4	25·5	47·5
Initial	1·2	23·4	3·4	25·5
Difference	22·2	22·0	22·1	22·0

Average accurate titre = 22·0 cm³

i.e., 22·0 cm³ HCl of unknown molarity reacted with 25·0 cm³ of 0·05M.Na$_2$CO$_3$. Hence if m is the molarity of the acid using the equation for the reaction

$$\frac{22 \times m}{25 \times 0.05} = \frac{2}{1}$$

$$m = 0.1136$$

The diluted hydrochloric acid is 0·114M. The molecular weight is 36·5, so the concentration of the acid is $0.1136 \times 36.5 = 41.5$ g/dm³.

If the dilution ratio is known accurately then the concentration of the original acid can be calculated.

If any other weight of sodium carbonate is taken then the molarity of the solution can be calculated from $\frac{4 \times \text{weight}}{106}$.

Experiment 2. *Alternative Method.* **Use of sodium metaborate to standardise a solution of hydrochloric acid.** (Approx. 0·1M hydrochloric acid; pure borax.)

It is best not to rely on one standardisation to determine the molarity of a solution.

Borax (Na$_2$B$_4$O$_7$,10H$_2$O) is not very soluble in cold water so it is used as a 0·05M solution.

$$Na_2B_4O_7, 10H_2O + 2HCl = 2NaCl + 4H_3BO_3 + 5H_2O$$

Boric acid
Too weak an acid to affect the methyl orange.

Molecular weight of borax = 381.

∴ For 250 cm³ 0·05M solution $381 \times 0.25 \times 0.05$ g of borax required

$$= 4.775 \text{ g}$$

Weigh out about 4·8 g of pure borax accurately, make up to 250 cm³. Take 25 cm³ of this solution, add a few drops of methyl orange and titrate with the 0·1M acid. The boric acid liberated does not affect the end-point if methyl orange is used as an indicator. This method can be used for nitric and hydrochloric acids, but with sulphuric acid the end-point is not satisfactorily defined.

24 THE ESSENTIALS OF VOLUMETRIC ANALYSIS

First calculate the molarity of the borax solution that has been made up, *i.e.*, $\dfrac{4 \times \text{weight}}{381}$ then using the titration results calculate the molarity of the acid.

Experiment 3. To standardise a solution of sodium hydroxide by means of oxalic acid. (Stick caustic soda; hydrated oxalic acid.)
N.B. Oxalic acid is very poisonous.

$2NaOH \quad + \quad H_2C_2O_4, 2H_2O^1 = Na_2C_2O_4 + 4H_2O$
Molecular weight 40 \qquad 126
For 250 cm^3 0·1M \qquad For 250 cm^3 0·05M solution use
solution use 1 g \qquad about 1·575 g

Quickly weigh out on to a watch-glass about 1·3 g of stick caustic soda (the purest form available for purchase). Put it into a beaker, add a little of water, swill round for a few moments and discard the solution. (This will have removed most of the surface layer of carbonate.) Dissolve the remainder in water and make up to about 250 cm^3 in a flask. Weigh out about 1·6 g of pure oxalic acid accurately in a weighing bottle. (Analytical quality should be used—oxalic acid frequently contains both calcium and potassium oxalates as impurity.) Dissolve the acid in water in a beaker and transfer the solution completely to a 250 cm^3 flask and make up to the mark, shake well or pour into a large flask and swirl. Fill a burette with the acid and titrate with 25 cm^3 of the alkali using phenolphthalein as an indicator (one drop should be sufficient to colour the solution pink). Run the acid in until the solution becomes colourless. Perform two or three accurate titrations.

First calculate the molarity of the oxalic acid solution that has been made up, *i.e.*, $\dfrac{4 \times \text{weight}}{126}$ then using the titration results calculate the molarity of the alkali.

Experiment 4. To standardise a solution of sodium hydroxide by means of succinic acid. (Caustic soda; succinic acid.)

Prepare a tenth molar solution of caustic soda as indicated in Experiment 3.

This hydrogen
does not ionise.
↓
$\begin{array}{l} CH_2 CO_2H \\ | \\ CH_2 CO_2H \end{array} + 2NaOH = (CH_2CO_2Na)_2 + 2H_2O$

[1] This water of crystallization takes no part in the reaction. It is included here because the crystals contain it.

ACIDIMETRY AND ALKALIMETRY

Molecular weight 118.

Hence weight of succinic acid for 250 cm^3 0·05M solution = 1·475 g

Weigh out accurately in a weighing bottle about 1·5 g of succinic acid and transfer it into a beaker. Warm in order to dissolve the acid. Cool the solution to room temperature and pour it together with the rinsings into a 250 cm^3 measuring flask. Make it up to the mark. To 25 cm^3 of the acid solution add 2 drops of phenolphthalein and run in the alkali from the burette, proceeding more cautiously as you are approaching the end-point. Shake well after each addition and note the reading at the first permanent pink tinge. Perform several accurate titrations.

First calculate the molarity of the succinic acid solution that has been made up, i.e., $\dfrac{4 \times \text{weight}}{118}$ then using the titration results calculate the molarity of the alkali.

N.B. Use a rubber stopper for the bottle in which you are keeping the caustic soda and wash the burette out with water and then with very dilute acid and again with water to prevent the tap of the burette from sticking in its socket. This procedure should be adopted whenever an alkaline solution is placed in the burette.

Experiment 5. Standardisation of an acid (HNO_3) by the Iceland Spar method (Concentrated nitric acid; Iceland Spar—a crystalline form of calcium carbonate).

This method depends on determining the weight of pure calcium carbonate dissolved by a known volume of the acid. It is suitable as a check, particularly when the acid solution is fairly concentrated. Note that it cannot be used to standardise sulphuric acid because calcium sulphate is only sparingly soluble.

$$CaCO_3 + 2HNO_3 = Ca(NO_3)_2 + H_2O + CO_2$$
100 g of calcium carbonate ≡ 2 dm^3 M.HNO_3
∴ 1·25 g of calcium carbonate ≡ 25 cm^3 M.HNO_3

Dilute some concentrated nitric acid by putting 25 cm^3 in about 125 cm^3 water in a measuring flask and then making the total volume up to 250 cm^3. Shake well. Weigh out accurately on a watch-glass a crystal of Iceland spar (2–3 g) and introduce it carefully into a conical flask and add about 50 cm^3 of water. Run into this water exactly 25 cm^3 of the nitric acid solution, place a funnel in the neck and leave in a safe place until the next day (or until effervescence ceases).

Pour away the solution surrounding the crystal and wash it several times with distilled water. Transfer the crystal to a watch-glass, dry it thoroughly in a steam-oven and weigh it.

Suppose loss in weight of crystal is 1·340 g.
From the equation 25 cm^3 M.HNO$_3$ ≡ 1·25 g of CaCO$_3$.

$$\therefore \text{Molarity of acid is } \frac{1\cdot340}{1\cdot25} \times 1$$

$$= 1\cdot072$$

Hence the molarity of the original acid is 10·7.

Experiment 6. To determine the molecular weight of calcium carbonate by the method of back titration. (Precipitated chalk; M. hydrochloric acid; 0·1M alkali.)

Calcium carbonate is insoluble in water and difficulties would arise in titrating a standard acid against it. These difficulties can be avoided by dissolving a known quantity of the carbonate in excess acid and determining the amount of excess by titration with alkali. The concentration of the latter should be less than that of the acid used, as the amount of acid left will be considerably less than the amount taken up.

Weigh accurately a weighing bottle containing about 1·5 g (not more) of pure dry precipitated calcium carbonate. Empty this carefully into a funnel resting in a 250 cm^3 flask containing 50 cm^3 M. hydrochloric acid and re-weigh the weighing bottle accurately. Wash the carbonate through into the flask and when effervescence has ceased make up to the mark with water and shake well. Withdraw 25 cm^3 and titrate against alkali. Use methyl orange or methyl red as an indicator.

$$CaCO_3 + 2HCl \rightarrow CaCl_2 + H_2O + CO_2$$

Calculation.

Suppose the weight of calcium carbonate taken = 1·47 g; and that 20·6 cm^3 0·1M alkali neutralised 25 cm^3 of residual solution.
∴ 20·6 cm^3 M alkali would neutralise 250 cm^3.
Volume of M acid originally present = 50 cm^3.
∴ (50 − 20·6) cm^3, *i.e.*, 29·4 cm^3 is the volume of M acid used up.
Hence 29·4 cm^3 M acid ≡ 1·47 g calcium carbonate.

$$\therefore 2000 \text{ cm}^3 \text{ M acid} \equiv \frac{1\cdot47}{29\cdot4} \times 2000 \text{ g of calcium carbonate.}$$

$$= 100 \text{ g}$$

∴ The molecular weight is 100.

In the determination of the molecular weight of an unknown compound on these lines it is necessary, before weighings are made, to determine the acidity of the base and hence the form of the equation. Then after making up the solution to 250 cm^3 it is necessary to verify that the solution is still acid, and if not, a further 25 cm^3

ACIDIMETRY AND ALKALIMETRY

of acid should be added. If, on back-titrating a 0·1M solution of alkali were found to be unsuitable a more concentrated solution could be used.

Experiment 7. To determine the number of molecules of water of crystallization in washing soda crystals. (0·2M hydrochloric acid; washing soda crystals.)

$$Na_2CO_3 + 2HCl = 2NaCl + H_2O + CO_2$$

Molecular weight 106 (of the anhydrous substance).

∴ 106 g of sodium carbonate ≡ 2 dm³ M acid.

Weight out accurately about 4 g of soda crystals (this is assuming about a third to be water of crystallization) being careful to pick translucent crystals. Transfer these to a 250 cm³ flask, add water, shaking after each addition, and make up to the mark. Shake well. Take 25 cm³ of this solution in a conical flask, add a few drops of methyl orange and run in the hydrochloric acid until the first permanent orange colour is observed. Repeat to obtain two or three accurate results.

Calculation.

Suppose the weight of soda crystals taken = 5·35 g.
Suppose 25 cm³ of soda solution required 18·7 cm³ 0·2M acid.

∴ Molarity of soda solution $= \dfrac{18·7}{25} \times \dfrac{0·2}{2}$

∴ Concentration in terms of anhydrous sodium carbonate $= \dfrac{18·7}{25} \times 0·1 \times 106 \text{ g/dm}^3$

∴ Weight of anhydrous sodium carbonate in 250 cm³ solution $= \dfrac{18·7 \times 0·1 \times 106}{25 \times 4} \text{ g}$

$= 1·982 \text{ g.}$

Let the formula be $Na_2CO_3.xH_2O$

then $\dfrac{Na_2CO_3}{Na_2CO_3,xH_2O} = \dfrac{1·982}{5·35} = \dfrac{106}{106 + 18x}$

Whence $x = 9·9$

∴ The number of molecules of water of crystallization = 10 (since x must be a whole number).

N.B. It is advisable to work with sodium carbonate solution in concentrations not less than 0·1M. In 0·05M solution the end-point using methyl orange is scarcely sharp enough.

Experiment 8. To determine the amounts of sodium carbonate and sodium hydroxide in a mixture. (*Double indicator method.*) (Mixture of about 5 g each, sodium hydroxide and sodium carbonate, in 1000 cm³ solution; 0·2M hydrochloric acid.)

THE ESSENTIALS OF VOLUMETRIC ANALYSIS

Principle.

		Indicator to exhibit completion of reaction.
I.	$NaOH + HCl = NaCl + H_2O$	Any indicator.
IIa.	$Na_2CO_3 + HCl = NaHCO_3 + HCl$	Phenolphthalein.
IIb.	$NaHCO_3 + HCl = NaCl + H_2O + CO_2$	Methyl orange.

As we have already seen the neutralisation of a strong alkali by means of a strong acid can be followed by the use of any indicator. Sodium hydrogen carbonate solution is slightly acid to phenolphthalein (you should verify this experimentally) whereas sodium carbonate solution is definitely alkaline to it. If acid is added to a mixture of sodium hydroxide and sodium carbonate in solution using phenolphthalein as indicator then the pink colour of the indicator is discharged when reactions I and IIa are complete. If methyl orange is now added and a further quantity of acid added the amount required will be that necessary to complete reaction IIb. But one molecule of sodium hydrogen carbonate has been formed from one molecule of sodium carbonate and hence the amounts of acid required for reactions IIa and IIb will be the same.

Suppose the volume of acid needed to reach the end-point as indicated by the phenolphthalein is a cm³ and the *additional* volume of acid to reach the end-point as indicated by the methyl orange is b cm³.

Then volume of acid reacting with sodium carbonate is $2 b$ cm³

,, ,, ,, ,, ,, hydroxide is
$(a + b) - 2 b$ cm³ $= (a - b)$ cm³

Method. Take 25 cm³ of the solution containing the two alkalis and add 1–2 drops of phenolphthalein solution and run in 0·2M acid until the pink colour is just discharged. Note the burette reading and add a few drops of methyl orange (or better methyl yellow) and add a further quantity of acid until the yellow colour of the methyl orange changes to orange.

Calculation. Suppose 25 cm³ of mixture required 21·3 cm³ of 0·2M HCl (phenolphthalein) and that the same 25 cm³ of mixture required a further 7·9 cm³ of 0·2M HCl (methyl orange).

Total volume of 0·2M HCl used = 29·2 cm³.

Half of the Na_2CO_3 required 7·9 cm³ of 0·2M acid.

∴ All the Na_2CO_3 required 15·8 cm³ of 0·2M acid.

∴ The NaOH required (29·2 − 15·8) cm³ or 13·4 cm³ of 0·2M acid.

∴ Concentration of the sodium carbonate solution is

$$\frac{15·8}{25} \times 0·2 \times \tfrac{1}{2} M$$

$$= \frac{15·8}{25} \times 0·1 \times 106 \text{ g/dm}^3$$

$$= 6·7 \text{ g/dm}^3.$$

ACIDIMETRY AND ALKALIMETRY

Concentration of the sodium hydroxide solution is $\dfrac{13\cdot 4}{25} \times 0\cdot 2$

$$= \dfrac{13\cdot 4}{25} \times 0\cdot 2 \times 40 \text{ g/dm}^3$$
$$= 4\cdot 29 \text{ g/dm}^3.$$

Note. The above exercise is a useful one on mixed indicators and one requiring thought. Unless the conditions are carefully controlled, however, and end-points correctly determined its accuracy leaves much to be desired. As an aid to the memory the student is reminded.

(*a*) the reaction between the alkali NaOH and hydrochloric acid cannot take place in two stages as can the reaction $Na_2CO_3 \to NaHCO_3 \to NaCl$ (see equations, p. 28);

(*b*) that the phenolphthalein must obviously be the indicator first added since it is colourless in acid solution and will not mask the colour of the methyl orange added later.

Experiment 9. Determination of the amounts of sodium carbonate and sodium hydroxide in a mixture (C. Winkler). (0·1M hydrochloric acid; mixture of 2–3 g each, sodium hydroxide and sodium carbonate in 1000 cm³ solution; barium chloride solution.)

Where accuracy is of prime importance the method due to Winkler is more satisfactory. The solutions used may be much more dilute without loss in accuracy.

Principle. The total alkali present (carbonate and hydroxide) is determined by titration with 0·1M acid using methyl orange as an indicator. To a second portion excess barium chloride solution is added

$$Na_2CO_3 + BaCl_2 = 2NaCl + BaCO_3 \downarrow$$

i.e., $\qquad Ba^{2+} + CO_3^{2-} \to BaCO_3 \downarrow$

Thus the carbonate is removed as insoluble barium carbonate. The hydroxyl ions from the alkali remain in solution. The concentration of these is determined in the usual way, using phenolphthalein as an indicator.

Method. Make up a solution of the mixture. To 25 cm³ of this solution add methyl orange and titrate with 0·1M hydrochloric acid until the yellow colour changes to orange. To a further 25 cm³ of the original solution add about an equal volume of 0·1M barium chloride solution and add 1–2 drops of phenolphthalein and titrate with 0·1M hydrochloric acid, noting the burette reading when the solution is decolourised. Take care to run in the acid slowly otherwise some of the barium carbonate may be acted upon before the end-point is reached.

Calculation. Suppose 25 cm³ of mixed sodium hydroxide and carbonate solution required a cm³ 0·1M HCl (methyl orange) and a further 25 cm³ after treatment with $BaCl_2$ required b cm³ 0·1M HCl (phenolphthalein).

∴ Volume of acid required to react with NaOH = b cm³
,, ,, ,, ,, ,, ,, Na_2CO_3 = $(a - b)$ cm³

∴ Concentration of NaOH $= \dfrac{b}{25} \times 0\cdot 1$ M

$= \dfrac{b \times 0\cdot 1 \times 40}{25}$ g/dm³

Concentration of Na_2CO_3 $= \dfrac{(a-b)}{25} \times 0\cdot 1 \times \tfrac{1}{2}$ M

$= \dfrac{(a-b) \times 0\cdot 1 \times 106}{25 \times 2}$ g/dm³

Experiment 10. Estimation of ammonia in ammonium sulphate by the indirect method. (M sodium hydroxide; 0·1M acid; ammonium sulphate.)

The ammonium salt (as solid or solution) is boiled with excess sodium hydroxide and the excess determined by back-titration with acid.

$(NH_4)_2SO_4 + 2NaOH = Na_2SO_4 + 2NH_3 + 2H_2O$
132 g   2 dm³M   34 g

∴ 1 g pure ammonium sulphate $\equiv \dfrac{2000}{132}$ cm³ M.NaOH

$\equiv 15\cdot 1$ cm³.

Weigh accurately a weighing bottle containing about 1·5 g of pure ammonium sulphate and transfer this to a conical flask. Reweigh the bottle. Run in exactly 50 cm³ of M.NaOH into the conical flask and insert in the flask a rubber stopper into which a reflux condenser tube is fitted—even a glass funnel will do, with care—this prevents loss of solution. Boil the contents of the flask for about 15 minutes or until no smell of ammonia is perceptible. Pour a little water down the condenser tube and so wash any alkali solution back into the flask. Pour the whole solution when cold into a 250 cm³ flask, adding several washings from the conical flask, and make up to the mark. Shake well. Titrate portions of 25 cm³ of this solution against the acid using methyl orange as an indicator.

Calculation. Suppose a g of ammonium sulphate were taken and that 25 cm³ of solution (made up to 250 cm³) required b cm³ of 0·1M acid.

∴ 250 cm³ would require b cm³ 0·1M acid.
∴ $(50 - b)$ cm³ of M.NaOH were used up.
But 2000 cm³M.NaOH ≡ 34 g NH_3.

∴ $(50 - b)$ cm³ M.NaOH ≡ $\dfrac{34}{2000} \times (50 - b)$ g NH₃

∴ % NH₃ in sample $= \dfrac{34}{2000} \dfrac{(50 - b)}{a} \times 100$

Experiment 11. Estimation of ammonia in an ammonium salt (e.g., ammonium sulphate) by the direct method. (0·5M sulphuric acid; 0·1M alkali; ammonium sulphate.)

Ammonium sulphate can be obtained in a very pure state, and this method is used in industry to standardise acid solutions.

Principle. A weighed amount of the ammonium sulphate is boiled with excess sodium hydroxide solution. The ammonia which is driven off is absorbed in excess of standard acid and the volume of the latter used up is found by back-titration.

Weigh accurately a weighing bottle containing about 1·5 g of ammonium sulphate (see Experiment 10) and empty this carefully into a funnel in a round-bottomed flask which should contain a few pieces of porous pot. Wash the sulphate through with distilled water. Reweigh the weighing bottle. Run in exactly 50 cm³ 0·5M sulphuric acid (or M hydrochloric acid) into a conical flask standing in cold water. Fit up the apparatus shown in Fig. 9. The trap T is to

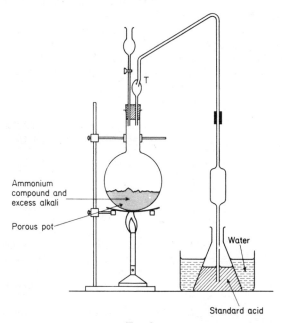

FIG. 9.

prevent droplets of alkali from being mechanically carried over into the standard acid.

Pour about 50 cm³ (a large excess) of 3M sodium hydroxide into the flask containing the ammonium sulphate. Warm the contents of the flask gently and finally boil for about ten minutes. At the end of that time open the tap and smell the vapour. If the ammonia can be detected the boiling must be continued for a few more minutes and the test repeated until no ammonia can be smelled. Remove the burner from underneath the solution, raise the pipette out of the acid solution, wash back into the conical flask any acid solution clinging to it inside or outside. Make up the residual acid to 250 cm³ taking care to include all washings and titrate 25 cm³ of this against 0·1M sodium hydroxide using methyl orange as an indicator. Calculate the percentage of ammonia from the fact that 1 dm³ $M.H_2SO_4 \equiv 34$ g NH_3.

Experiment 12. To determine the degree of temporary hardness in water. (0·1M hydrochloric acid; temporarily hard water.)

Prepare a temporarily hard water (if one is not available) by bubbling a rapid stream of carbon dioxide through calcium hydroxide solution (there is no need to wait until the solution is clear—any precipitate can be filtered off). Take 100 cm³ of the solution, add a few drops of methyl orange and run in the standard hydrochloric acid until the yellow colour changes to orange.

$$Ca(HCO_3)_2 + 2HCl = CaCl_2 + 2H_2O + 2CO_2$$
$$162 \text{ g} \qquad 2 \text{ dm}^3 M$$

Suppose a cm³ of 0·104M acid are required for 100 cm³ of the temporarily hard water then the weight of $Ca(HCO_3)_2$ in 100 cm³ solution is $\dfrac{162 \times a \times 0·104}{2000 \times 1}$ g.

But 162 g of calcium bicarbonate would form 100 g calcium carbonate if heated so the weight of calcium carbonate from 100 cm³ solution is $\dfrac{100 \times a \times 0·104}{2000 \times 1}$ g. Hence the hardness can be calculated in parts per million.

Exercises

An asterisk indicates that the particular problem can be attacked without the previous preparation of solutions, the concentrations of which have not to be divulged to the student.

(1) [Formaldehyde solution, 4% solution; M. ammonium chloride; M. sodium hydroxide.]

Investigate the reaction between ammonia and formaldehyde in aqueous solution. The following experiments are suggested.

ACIDIMETRY AND ALKALIMETRY

(a) Mix excess (50 cm^3) of the formaldehyde solution with 20 cm^3 of the ammonium chloride; add 40 cm^3 of the sodium hydroxide, and leave the liquid fifteen minutes before titrating with hydrochloric acid using phenolphthalein.

(b) Mix excess (50 cm^3) of the ammonium chloride with 20 cm^3 of the formaldehyde solution, add 40 cm^3 of the sodium hydroxide, and leave for fifteen minutes before titrating with molar hydrochloric acid. Litmus may be used, but bromo-thymol blue will be found more satisfactory. Repeat the experiment but omit the formaldehyde. State briefly the conclusions at which you arrive. (Cambridge Schol.)

(2)* [M. ammonia; 0·1M copper sulphate; 0·025M sulphuric acid; chloroform.]

Determine the composition of the complex tetramminecopper(II) ion formed when excess ammonia is added to a copper sulphate solution. Mix 25 cm^3 of M. ammonia with 25 cm^3 of 0·1M copper sulphate. Shake the resulting liquid with 75 cm^3 of chloroform for ten minutes. Separate off 50 cm^3 of the chloroform and determine the amount of ammonia in it by titration in a stoppered bottle with the standard acid provided. Assume that free ammonia distributes itself between the solution and chloroform in the ratio $\dfrac{\text{Concentration on aqueous solution}}{\text{Concentration in chloroform}} = 26$

Note. Use a measuring cylinder and not a pipette for the chloroform.
(Cambridge Schol.)

(3) [Impure calcium carbonate; M. hydrochloric acid; 0·1M sodium hydroxide.]

The substance A is a mixture of calcium carbonate and silicon dioxide. Find its percentage composition by weight.

Weigh out accurately about 1 g of the mixture and transfer it to a graduated 250 cm^3 flask. Add 50 cm^3 of hydrochloric acid. When the reaction is complete make the liquid up to the mark with distilled water, shake well to ensure thorough mixing and titrate 25 cm^3 portions with sodium hydroxide solution using methyl orange as indicator. (J.)

(4)* [Sodium hydrogen sulphate; sodium hydrogen carbonate.]
Determine the purity of the laboratory specimen of sodium hydrogen sulphate. You are supplied with pure sodium carbonate.

(5) [Sodium hydroxide and ammonium chloride solution; 0·1M hydrochloric acid.]

You are provided with a solution containing ammonium chloride and sodium hydroxide in such proportions that there is less than 1 mole of ammonium chloride for every one of sodium hydroxide. Find the concentration of each in g/dm^3 using the standard acid.

(6)* [0·05M sodium carbonate.]
Estimate the concentration of the barium chloride solution in the reagent bottles. You are supplied with standard sodium carbonate solution. Phenolphthalein will show the end-point of the reaction:—

$$BaCl_2 + Na_2CO_3 = BaCO_3 \downarrow + 2NaCl$$

(7) *Determination of Distribution Ratio of Succinic Acid between Water*

and Ether. [Succinic acid, ether, 0·5M sodium hydroxide; 0·1M sodium hydroxide.]

Grind a little succinic acid in a mortar and weigh out quantities of 1·5 g, 1·0 g and 0·5 g (these weighings need only be roughly done). Pour into a separating funnel about 40 cm³ of water and add the 1 g of succinic acid. Add about 40 cm³ of ether, shake until the acid has dissolved and allow to stand for a few minutes. Run off nearly all the water layer and titrate 25 cm³ of this against 0·5M NaOH using phenolphthalein as an indicator (about 30 cm³ will be required). Discard the boundary portion. Take 25 cm³ of the ether layer by means of a pipette filler (fairly accurate results can be obtained by using a measuring cylinder) and titrate this against the 0·1M NaOH. Shake well after each addition of alkali. (The volume of the 0·1M solution required will be approximately the same as the amount of 0·5M solution required for the water layer.)

Obtain the distribution ratio:—

$$K = \frac{5 \times [\text{Vol. of 0·5M NaOH for 25 cm}^3 \text{ aqueous layer}]}{\text{Vol. of 0·1M NaOH for 25 cm}^3 \text{ ether layer}}.$$

Repeat the process with the other quantities of the succinic acid.

(8)* *Quantitative Verification of Solubility Product* (F. G. Mee). [Pure sodium hydroxide 0·1M; 0·1M hydrochloric acid.

Calcium hydroxide should have a smaller solubility in a solution of sodium hydroxide than in water, and it is the object of this experiment to verify that the decrease of solubility that occurs is that which would be expected from the equilibrium constant for the electrolytic dissociation:—

$$Ca(OH)_2 \rightleftharpoons Ca^{2+} + 2OH^-$$

Extra OH^- are provided from some other source—sodium hydroxide in this case—the concentration of Ca^{2+} ions decreases in such a way that the product of it and the square of the concentration of OH^- remains a constant, if the solution is saturated.

To verify this solutions of calcium hydroxide are made up in (1) water, (2) 0·025M NaOH, (3) 0·05M NaOH, (4) 0·1M NaOH. The alkali should be as free as possible from carbonate and the solutions can be made by adding a quantity of calcium hydroxide to each of the sodium hydroxide solutions and allowing to stand, with shaking at intervals. Each of these solutions is titrated with 0·1M HCl, which will, in each case, give a measure of the total OH^- concentration present. The OH^- concentration due to the sodium hydroxide is known, assuming this to be fully ionised also. Subtraction will yield the concentration of OH^- due to dissolved calcium hydroxide, and the concentration Ca^{2+} is exactly half of this. Titration thus gives all required to calculate the solubility product.

A typical set of results is the following:—

25 cm³ of each of the four solutions were titrated with 0·1M HCl, using phenolphthalein indicator, with results shown in the second column of the table shown on opposite page.

The constancy of the product is striking. The effect of decreasing the solubility due to the sodium hydroxide appears during the experiment, before the detailed calculations is made. Calcium hydroxide in water gave a titration 11·9 cm³; 0·05M NaOH alone would give 12·5 cm³. Yet calcium hydroxide in 0·05M NaOH gives but 17·7 cm³, showing that the solubility has been reduced to about one-half. The result with the 0·1M

ACIDIMETRY AND ALKALIMETRY 35

Solution in	Titration cm³	Ionic concentration in mole/dm³				$[Ca^{2+}] \times [OH^-]^2$ ($\times 10^5$)
		Total OH⁻ present	OH⁻ due to soda	OH⁻ due to lime	Ca^{2+}	
Water	11·4	0·048	—	0·048	0·0237	5·4
0·025M NaOH	14·5	0·058	0·025	0·033	0·0165	5·6
0·05M NaOH	17·7	0·071	0·050	0·021	0·0105	5·3
0·1M NaOH	27·3	0·109	0·100	0·009	0·0045	5·4

solution cannot be expected to be very accurate, since at this concentration the calcium hydroxide contributes so very little to the total hydroxide ion concentration. If more than four results are desired, they should therefore be taken with sodium hydroxide solutions between the above values, and not more concentrated than 0·1M.
(From the Science Masters' Book published by Messrs. John Murray.)

Questions and Calculations

(1) Explain the term *molar* solution.
 What weights of potassium hydroxide, sulphuric acid, sodium carbonate are present respectively in a 1000 cm³ of a tenth molar solution of each substance?
 Describe how you would find the concentration of a solution of sodium carbonate if you were provided with 0·05M solution of sulphuric acid.
(2) 0·56 g of the oxide of a divalent metal was dissolved in 250 cm³ of a solution of hydrochloric acid containing 3·65 g per 1000 cm³. To neutralise the excess of acid 50 cm³ of a solution of sodium hydroxide containing 4·0 g/dm³ were required. Calculate the atomic weight of the metal. (J.)
(3) What do you understand by the terms acid, base, salt? 10cm³ of a solution of potassium hydroxide containing 4 g/dm³ of KOH required for neutralisation 12·5 cm³ of a solution of sulphuric acid.
 Calculate the number of grammes of sulphuric acid in 1000 cm³ of solution. What is the molarity of the solution?
(4) 4·35 g of a mixture of sodium chloride and anhydrous sodium carbonate were dissolved in distilled water and the solution made up to 100 cm³. 20 cm³ of this solution required 75 cm³ of 0·05M sulphuric acid to react completely with the sodium carbonate. Calculate the percentage composition by weight of the original mixture.
(J.)
(5) A solution of sodium carbonate and sodium hydrogen carbonate gave the following results when analysed. Explain the two methods adopted, and calculate the weight of each salt present in 1 dm³ of the solution.

(a) 20 cm³ required 11·1 cm³ of 0·09 M HCl for neutralisation when phenolphthalein was used as the indicator, and a further 33·3 cm³ of acid when methyl orange was added.

(b) 20 cm³ of 1·1M NaOH were added to 100 cm³ of the solution. Excess of barium chloride solution was then added, and the liquid was finally titrated with 0·8M HCl, the indicator used being phenolphthalein, 15 cm³ were required. (Use also the result of the methyl orange titration given in (a).) (O.)

4. Potassium Permanganate

POTASSIUM permanganate, $KMnO_4$, is a powerful oxidising agent and is used for the estimation of many reducing agents, especially compounds of iron, and oxalic acid and its salts.

Condition of Use of Potassium Permanganate. In acid solution two molecules of potassium permanganate react with, for example, five molecules of iron(II) sulphate in the presence of dilute sulphuric acid.

$$2KMnO_4 + 8H_2SO_4 + 10FeSO_4 \rightarrow K_2SO_4 + 2MnSO_4 + 5Fe_2(SO_4)_3 + 8H_2O$$

or, in ionic terms

$$2MnO_4^- + 16H^+ + 10Fe^{2+} \rightarrow 2Mn^{2+} + 10Fe^{3+} + 8H_2O.$$

These equations can be considered from the point of view of the oxidising agent, the potassium permanganate, which gains electrons to be

$$MnO_4^- + 8H^+ + 5e^- \rightarrow Mn^{2+} + 4H_2O$$

and of the reducing agent, the iron(II) salt, which supplies the electrons:

$$Fe^{2+} \rightarrow Fe^{3+} + e^-$$

In alkaline solution, potassium permanganate by a different reaction yields manganese dioxide as a brown precipitate. Consideration of these facts makes it clear at once that for quantitative work, potassium permanganate must be used in conditions which exclude entirely one of these reactions. In practice, potassium permanganate is almost always used to titrate solutions which are sufficiently acidic to exclude altogether the formation of manganese dioxide.

Of the three mineral acids available, only sulphuric acid is suitable for use with potassium permanganate, for this compound reacts with hydrochloric acid:—

$$2KMnO_4 + 16HCl = 2KCl + 2MnCl_2 + 8H_2O + 5Cl_2,$$

while nitric acid is itself an oxidising agent and might interfere with the oxidising action of the permanganate. The solution which is in process of titration with potassium permanganate must be sufficiently acidic to prevent the formation of any precipitate of manganese dioxide. As bench sulphuric acid is usually 1·5M and potassium permanganate solution about 0·02M, a bulk of bench acid equal to a third that of the solution to be titrated will usually provide a sufficient excess of acid.

Indicator and End-point. As the titration proceeds, manganese ions accumulate, but at the dilution used, give a colourless solution.

As soon as potassium permanganate is in excess, the solution becomes pink and therefore potassium permanganate acts as its own indicator, the end-point being the first permanent pink coloration.

Standard Solution of Potassium Permanganate. The equation derived in the preceding paragraph is

$$MnO_4^- + 8H^+ + 5e^- \rightarrow Mn^{2+} + 4H_2O$$

solutions of potassium permanganate are therefore usually made having a concentration of one fifth of one tenth molar, *i.e.*, 0·02M, 1 dm^3 of 0·02M.KMnO$_4$ solution contains 3·16 g of the salt.

An accurately 0·02M solution of potassium permanganate cannot be made up directly from the solid because this may be reduced by organic matter from the atmosphere and so rendered impure further, organic matter present in the water in which the salt is dissolved may reduce it. It is therefore desirable to make up a solution slightly more concentrated (say about 3·25 g/dm^3) and allow it to stand several days. It may then be standardised by methods described below. A potassium permanganate solution slowly decomposes and should be protected from light and standardised again at intervals. It may be standardised by a pure iron(II) salt or a pure oxalate.

Experiment 13. Standardisation of potassium permanganate solution by an iron(II) salt (iron(II) ammonium sulphate). Approx. 0·02M KMnO$_4$; (iron(II) ammonium sulphate).

The oxidation of an iron(II) salt by potassium permanganate may be expressed ionically thus:—

$$Fe^{2+} \rightarrow Fe^{3+} + e^-$$

or using iron(II) sulphate, the oxidation may be written in molecular terms:—

$$10FeSO_4 + 2KMnO_4 + 8H_2SO_4 = 5Fe_2(SO_4)_3 + K_2SO_4 + 2MnSO_4 + 8H_2O$$

Iron(II) sulphate crystals, FeSO$_4$, 7H$_2$O, cannot be used for standardisation because they are rendered impure by efflorescence and by atmospheric oxidation to form a brown basic sulphate as a result of a reaction of the type:—

$$12FeSO_4 + 3O_2 + 6H_2O = 4[Fe(OH)_3, Fe_2(SO_4)_3].$$

The salt, iron(II) ammonium sulphate, FeSO$_4$, (NH$_4$)$_2$SO$_4$,6H$_2$O, is free from these disadvantages and can be obtained in a high state of purity. (It is prepared by dissolving iron(II) and ammonium sulphates in the calculated quantities in hot water containing sulphuric acid and allowing the solution to crystallize.) In solution, it breaks up into iron(II) ions, sulphate ions and ammonium ions. Only the former react with the permanganate.

One molecule of iron(II) ammonium sulphate contains one iron(II) ion and the partial equation given above shows that this loses

POTASSIUM PERMANGANATE

one electron, thus the weight of the salt to be taken for 250 cm³ 0·1M solution is 0·25 × 0·1 × 392 g, *i.e.*, 9·8 g.

This amount should be weighed out in a weighing bottle. To prevent oxidation, the salt should now be dissolved in diluted sulphuric acid, which has been boiled to remove air and then cooled. Make up 250 cm³ of solution in this way: it is then accurately 0·1M.

Titration. Measure out 25 cm³ of the solution into a conical flask. Add about 10 cm³ of bench sulphuric acid and titrate with the potassium permanganate solution from a burette until the first permanent pink coloration is observed. No brown precipitate should appear. Repeat the titration twice.

Suppose that 25 cm³ iron(II) solution is oxidised by an average volume of 23·4 cm³ potassium permanganate solution. Then, since the iron(II) salt solution is exactly 0·1M the molarity of the permanganate solution must be $\frac{25 \times 0·1}{23·4 \times 5}$ or 0·02136, and hence $KMnO_4$ contains 0·02136 × 158 = 3·375 g/dm³.

Experiment 14. Standardisation of potassium permanganate solution by sodium oxalate $Na_2C_2O_4$. (Approx. 0·02M potassium permanganate; sodium oxalate.)

An acidified solution of an oxalate is for purposes of titration with potassium permanganate solution equivalent to a solution of oxalic acid itself:—

$$Na_2C_2O_4 \rightleftharpoons 2Na^+ + C_2O_4^{2-}$$
$$C_2O_4^{2-} + 2H^+ \rightleftharpoons H_2C_2O_4$$
from acid

Sodium oxalate is used for standardisation because it can be obtained in a pure state more easily than can oxalic acid.

The oxidation of an oxalate is represented essentially by the equation:—

$$C_2O_4^{2-} \rightarrow 2CO_2 + 2e^-$$

The molecular equation is:—

$$2KMnO_4 + 5H_2C_2O_4 + 3H_2SO_4 = K_2SO_4 + 2MnSO_4 + 8H_2O + 10CO_2$$

From these 250 cm³ 0·05M $Na_2C_2O_4$ solution will therefore contain 0·25 × 0·05 × 134 = 1·675 g of the salt.

Weigh out this amount and make up to 250 cm³ of solution with distilled water. When measuring this solution in a pipette be careful not to allow it to enter your mouth—use a pipette filler. Oxalates are poisonous.

Titration. Potassium permanganate does not oxidise oxalates in cold solution; a temperature of 70°C is necessary to cause the reaction to begin.

To 25 cm³ of the 0·05M $Na_2C_2O_4$ solution in a conical flask add about 10 cm³ of bench sulphuric acid and heat the mixture to 70°C. This temperature can be estimated accurately enough by testing with the palm of the hand. When the bottom of the flask is just too hot to hold, the temperature of the liquid is approximately correct. Titrate with potassium permanganate, heating again as the liquid cools, till a permanent pink coloration is observed. Manganese sulphate formed during the reaction has a catalytic effect, but side reactions are prevented if the temperature is still maintained at above 60°C. Repeat the titration twice with further portions of 25 cm^3.

Calculate the concentration of the permanganate solution.

Experiment 15. Determination of the number of molecules of water of crystallization in a molecule of iron(II) sulphate crystals ($FeSO_4, xH_2O$). (0·02M potassium permanganate; iron(II) sulphate crystals.)

$$10(FeSO_4, xH_2O) + 2KMnO_4 + 8H_2SO_4 = 5Fe_2(SO_4)_3 + K_2SO_4$$
Molecular weight $152 + 18x \qquad + 2MnSO_4 + (10x + 8)H_2O.$

For 250 cm^3 0·1M solution use $(152 + 18x) \times 0.25 \times 0.1$ g of iron(II) sulphate crystals, i.e., $(3.8 + 0.45x)$ g.

x will probably be between 1 and 10, i.e., the weight of crystals should be between 4·3 g and 8·3 g. Weigh out about 7 g of crystals and make up to 250 cm^3 of acidified solution, as described for iron(II) ammonium sulphate (p. 38). Titrate against 0·02M potassium permanganate solution.

Calculation. Suppose a g of crystals were made up to 250 cm^3 of solution. Hence the molarity (m) and weight of crystals in the solution.

$$a = (3.8 + 0.45x)\frac{m}{0.1}.$$

Solve for x.

Experiment 16. Estimation of the percentage by weight of iron in iron wire. (0·02M potassium permanganate; iron wire; flask and Bunsen valve.)

A suitable weight of iron is converted to iron(II) sulphate solution, which is then titrated with potassium permanganate.

To obtain 250 cm^3 0·1M iron(II) sulphate solution $56 \times 0.25 \times 0.1$ g pure iron (i.e., 1·4 g) would be required. Weigh accurately 1·3 to 1·5 g of the wire.

Preparation of the Iron(II) Sulphate Solution. The iron is treated with dilute sulphuric acid, heat being necessary to secure a sufficiently rapid reaction. Conditions must be such that iron(III) compounds are not formed. While the reaction is actually proceeding, the hydrogen generated acts as a reducing agent, but when solution is complete and the liquid is still hot, entry of air might bring about oxidation. To prevent this, the "Bunsen valve" is used. It consists of a piece of rubber tubing, carrying a longitudinal slit for part of its length,

POTASSIUM PERMANGANATE

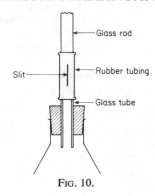

Fig. 10.

closed at the upper end by glass rod and connected by a stopper and glass tubing to the reaction flask (Fig. 10). The slit opens outwards only; it allows the escape of hydrogen, but when the flask is cooling air cannot be drawn in because the valve closes. When the solution is cool, the risk of oxidation is negligible.

Fit a "Bunsen valve" to a small flask and in it warm the weighed iron wire with dilute sulphuric acid till all the iron is dissolved.

$$Fe + H_2SO_4 = FeSO_4 + H_2$$

The solution will probably be cloudy. This effect is caused by precipitation of fine particles of carbon, an impurity in the iron. They will not interfere. When the reaction is complete and the flask has cooled, transfer the solution to a 250 cm^3 flask, wash out the reaction flask several times with air-free distilled water and add the washings to the 250 cm^3 flask. Make up to the mark with air-free distilled water and shake well. Titrate against the potassium permanganate.

Calculation. Calculate the molarity (m) of the iron(II) solution and hence the weight of iron. From the latter the percentage purity can be evaluated.

Experiment 17.   Estimation of iron in iron(III) ammonium sulphate.
(0·02M potassium permanganate; iron(III) ammonium sulphate zinc.)

Estimation of iron in the iron(III) state is carried out by first reducing it quantitatively to the iron(II) state and then titrating the resulting iron(II) solution with potassium permanganate. The molecular weight of the alum is 482, corresponding to a formula $Fe(NH_4)(SO_4)_2, 12H_2O$. Thus for iron(III) ammonium sulphate solution $482 \times 0.25 \times 0.1$ g, *i.e.*, 12·05 g of alum are required. Weigh out about 12 g of the alum and make up this weight to 250 cm^3 of solution, acidifying with dilute sulphuric acid to prevent hydrolysis.

Reduction of the Iron(III) Salt to the Iron(II) State. Measure out 25 cm³ of the alum solution in a conical flask, add dilute sulphuric acid and several pieces of zinc. (The zinc must be free from iron.) Allow effervescence to proceed for twenty minutes (set up two more similar reduction flasks in the meantime). The mixture may then be tested for completion of the reduction.

$$Zn + 2Fe^{3+} \rightarrow Zn^{2+} + 2Fe^{2+}$$

Test. Place on a white tile a drop of solution of potassium thiocyanate, KSCN. Dip a glass rod into the reduction flask and allow a drop of liquid from it to mix with the drop of solution on the tile. If a reddish brown tinge appears, an iron(III) salt still remains and reduction is not complete.

$$Fe^{3+} + CNS^- = FeSCN^{2+}$$

Continue the reduction until the test is negative, *i.e.*, the mixture remains colourless or is only very faintly coloured.

Cool the contents of the flask, filter through glass-wool (for speed), wash the zinc and flask well with air-free distilled water. Add the washings to the main solution and titrate against potassium permanganate. (Acidify further if necessary.) Repeat with the other two flasks, but delay the test for completion until reduction has proceeded for at least as long as was necessary before. This prevents loss of solution in repeated testing.

Calculation. Calculate the molarity and weight of iron in the alum solution. From the latter result the percentage of iron in the alum can be calculated.

Experiment 18. Other methods of reducing the iron(III) salt: use of zinc amalgam.

The method of reduction given above is rather slow. A more rapid result is achieved by using zinc amalgam. To prepare it, weigh out 200 g of mercury and 5 g of zinc foil. Put the mercury into a dish and heat it on the steam bath. Add a few drops of dilute sulphuric acid and stir into the mercury a thin strip of the zinc foil until all the zinc is dissolved. If the amalgam solidifies when cold, warm it and add a little more mercury. Add a small quantity of dilute sulphuric acid to the amalgam in a stoppered conical flask. Run in 25 cm³ of the alum solution, insert the stopper and shake gently for a few minutes. No gas is evolved. The reduction should be complete in about five minutes. Decant off the reduced solution (test as described in Expt. 17) into a flask, wash the amalgam several times with small quantities of distilled water, adding the washings to the reduced solution. Titrate against potassium permanganate. The amalgam may now be used to reduce a further portion of 25 cm³ of the alum solution.

POTASSIUM PERMANGANATE

Experiment 19. Estimation of oxalic acid and one of its soluble salts in a mixture of the two, e.g., oxalic acid and sodium oxalate. (0·02M potassium permanganate; 0·1M sodium hydroxide; mixture of oxalic acid and sodium oxalate.)

The oxalic acid is determined separately by titration with 0·1M NaOH; the total oxalate is determined by titration with 0·02M $KMnO_4$.

$$H_2C_2O_4 + 2NaOH = 2H_2O + Na_2C_2O_4$$

Molecular weight 90.

For 250 cm³ 0·05M solution $90 \times 0.250 \times 0.05$ g of anhydrous oxalic acid are needed, i.e., 1·125 g. Allowing for the fact that the given mixture will probably contain hydrated crystals of the acid ($H_2C_2O_4,2H_2O$) and for the higher molecular weight of sodium oxalate ($Na_2C_2O_4 = 134$) about 2 g of the mixture will probably provide 250 cm³ of a suitable solution. Weigh out accurately about 2 g of the mixture and make up to 250 cm³ of solution.

Titrations. (a) Titrate 25 cm³ of the solution with 0·1M NaOH using phenolphthalein as indicator.

(b) Titrate 25 cm³ of the solution, acidified with dilute sulphuric acid and heated to 70°C with 0·02M $KMnO_4$.

The sodium hydroxide solution reacted with the free oxalic acid only; the potassium permanganate solution reacted with the oxalic acid and sodium oxalate.

But since both oxalic acid and sodium oxalate contain one oxalate ion in a molecule the volume of potassium permanganate can be separated into the portions reacting with oxalic acid and sodium oxalate.

Calculation. Suppose a g of the mixture were made up to 250 cm³ solution and 25 cm³ of this needed b cm³ 0·1M NaOH and d cm³ 0·02M $KMnO_4$ for titration.

Calculate the molarity and hence the weight of anhydrous oxalic acid in 250 cm³ solution:

Wt. of oxalic acid in 250 cm³ solution $= \dfrac{90 \times 10b}{20\,000}$ g

∴ Per cent of oxalic acid in the mixture $= \dfrac{90 \times 10b}{20\,000} \times \dfrac{100}{a}$.

$2MnO_4^- \equiv 5C_2O_4^{2-}$
100 dm³ 0·02M 5×134 g of sodium oxalate.

Calculate the molarity of the solution with respect to oxalate ions and subtract the molarity of oxalic acid. Hence calculate the weight of sodium oxalate in 250 cm³ solution and then its percentage.

Experiment 20. Estimation of hydrogen peroxide. ("10- or 20-vol." hydrogen peroxide; 0·02M potassium permanganate.)

An acidified solution of hydrogen peroxide reacts with potassium permanganate,[1] liberating oxygen, and can be estimated by this reaction.

$$5H_2O_2 + 2KMnO_4 + 3H_2SO_4 = K_2SO_4 + 2MnSO_4 + 8H_2O + 5O_2.$$
$$H_2O_2 \rightarrow O_2 + 2H^+ + 2e^-.$$

The "10-volume" and "20-volume" solutions of hydrogen peroxide are both more concentrated than 0·05M, but, as hydrogen peroxide decomposes slowly at ordinary temperature, the appropriate ratio of dilution for a given sample must be found by trial.

Estimation of Dilution Ratio. By means of a pipette measure out 1 cm³ of the given solution of hydrogen peroxide into a conical flask, acidify it with dilute sulphuric acid and run in potassium permanganate from a burette 0·5 cm³ at a time till the mixture is pink. No attempt should be made to obtain an accurate end-point. If d cm³ of potassium permanganate solution are needed, the hydrogen peroxide solution must be diluted with distilled water so that the original and diluted volumes are in the ratio of 1 : d approximately.

Titration. Repeat the titration with 25 cm³ of the diluted hydrogen peroxide solution, acidify with dilute acid and obtain an accurate end-point.

Calculation. Suppose a cm³ of the original hydrogen peroxide solution were diluted to b cm³ and 25 cm³ of the diluted solution were titrated by p cm³ 0·02M KMnO$_4$.

$$2KMnO_4 \equiv 5H_2O_2$$
$$100 \text{ dm}^3 \text{ } 0\cdot02M \quad 5 \times 34 \text{ g}$$

Weight of hydrogen peroxide in 25 cm³ diluted solution

$$= (5 \times 34) \times \frac{p}{100\,000} \text{ g}$$

∴ Weight of hydrogen peroxide in diluted solution

$$= (5 \times 34) \times \frac{p}{100\,000} \times \frac{1000}{25} \text{ g/dm}^3$$

and hence the concentration of the original solution.

Experiment 21. Estimation of percentage of purity of commercial sodium nitrite. (Sodium nitrite; 0·02M potassium permanganate.)

An acidified solution of sodium nitrite is oxidised by potassium permanganate to nitrate.

$5NaNO_2 + 2KMnO_4 + 3H_2SO_4 = 5NaNO_3 + K_2SO_4 + 2MnSO_4$
Molecular weight 69 $\hfill + 3H_2O$
$$NO_2^- + H_2O \rightarrow NO_3^- + 2H^+ + 2e^-$$

The nitrite solution cannot be titrated with permanganate solution

---

[1] Mannitol and other carbohydrates are often present in hydrogen peroxide solution to retard its decomposition; they will react giving a high reading.

POTASSIUM PERMANGANATE 45

from a burette in the usual way, because, as soon as it is acidified, the nitrous acid formed begins to decompose.

$$2NaNO_2 + H_2SO_4 = Na_2SO_4 + 2HNO_2$$
$$3HNO_2 = HNO_3 + H_2O + 2NO.$$

The nitrite solution is placed in the burette and is added, *slowly* and with constant stirring, to an acidified solution of potassium permanganate.

For 250 cm^3 0·05M sodium nitrite solution, $69 \times 0.25 \times 0.05$ or 0·86 g would be required.

To allow for impurity make up a solution containing about 1 g of sodium nitrite in 250 cm^3.

Titration. Pipette 25 cm^3 0·02M potassium permanganate solution into a beaker, acidify with 10 cm^3 of bench dilute sulphuric acid and, from a burette add the sodium nitrite solution slowly and with continual stirring till the permanganate colour is just discharged. Repeat the titration to obtain two concordant results.

Calculation. Calculate the molarity and then the weight of pure sodium nitrate used.

Hence the percentage purity of sodium nitrite

$$= (5 \times 69) \times \frac{25}{100\,000} \times \frac{250}{b} \times \frac{100}{a}.$$

Experiment 22. Given that iron(II) ammonium sulphate has the formula FeSO$_4$.(NH$_4$)$_2$SO$_4$,xH$_2$O determine x. (Iron(II) ammonium sulphate; 0·02M potassium permanganate.)

$$10[FeSO_4.(NH_4)_2SO_4, xH_2O] \equiv 2KMnO_4.$$

Making a reasonable assumption for a value of x calculate the approximate weight of iron(II) ammonium sulphate solution needed for 250 cm^3 0·1M solution. Make up this solution in dilute sulphuric acid (p. 38) and titrate with potassium permanganate.

Experiment 23. Analyse a given mixture of potassium sulphate and potassium permanganate. (Mixture of potassium sulphate and potassium permanganate; 0·1M iron(II) ammonium sulphate.)

Make up the standard solution of iron(II) ammonium sulphate (p. 38) in dilute sulphuric acid. Assuming that the proportion of potassium permanganate in the given mixture is about 50%, make up a solution of accurately known concentration and about 0·02M. Titrate the iron(II) ammonium sulphate solution with it.

$$10[FeSO_4.(NH_4)_2SO_4, 6H_2O] \equiv 2KMnO_4$$

Experiment 24. Determine the solubility of ammonium oxalate in water at room temperature in g/dm^3 of solution of anhydrous salt. (Ammonium oxalate, 0·02M KMnO$_4$.)

Prepare about 100 cm³ of a saturated solution of ammonium oxalate at room temperature by heating the salt with distilled water at about 50°C, adding so much of the salt that a small sample cooled in a test tube yields crystals. Cool the whole solution under the tap, determine its temperature and filter through dry glass-wool into a dry flask. Titrate 1 cm³ of this filtrate, using the method of Expt. 20, with the potassium permanganate. On this evidence, dilute a known volume of the saturated solution to 250 cm³ in a measuring flask to give approximately a 0·05M solution.

Titrate portions of 25 cm³ with 0·02M $KMnO_4$ solution:

$$5(NH_4)_2C_2O_4 \equiv 2KMnO_4 \text{ (see p. 39)}$$
$$5 \times 124 \text{ g} \equiv 100 \text{ dm}^3 \text{ } 0·02M.$$

Experiment 25. A study of potassium hydrogen oxalate. (2 g/250 cm³ "tetroxalate" solution, 0·1M sodium hydroxide, 0·02M potassium permanganate.)

Caution—oxalates are poisonous.

Titrate 25 cm³ portions of the oxalate solution with the alkali using phenolphthalein as the indicator. Then take a new 25 cm³ portion of the oxalate solution, add 100 cm³ of bench dilute sulphuric acid and after warming the solution to 70°C titrate rapidly with the permanganate.

If the formula of the oxalate is $K_aH_b(C_2O_4)_c,dH_2O$ calculate a, b, c and d.

Experiment 26. Determine the percentage of manganese(IV) oxide in a given sample of pyrolusite. (Pyrolusite; 0·25M oxalic acid; 0·02M potassium permanganate.) (See Expt. 29 for alternative method.)

Assume that manganese(IV) oxide reacts with oxalic acid in sulphuric acid solution according to the equation:—

$$MnO_2 + H_2C_2O_4 + H_2SO_4 = MnSO_4 + 2H_2O + 2CO_2$$

Make up 250 cm³ of a solution of oxalic acid which is about 0·25M, i.e., 31·5 g/dm³ of crystals.

Transfer 100 cm³ of it (accurately measured by pipette, using a pipette filler) to a flask and add 20 cm³ bench sulphuric acid. Weigh a weighing-bottle containing about 1·1 g of the powdered pyrolusite, add the powder to the acidified oxalic acid solution and obtain the weight of the powder by difference. Boil the mixture gently (with a funnel in the neck of the flask) until the remaining solid particles (silica) are white. Transfer the liquid after cooling to a 250 cm³ flask, wash the flask and funnel and add the washings to the bulk of the solution. Make up to the mark. Titrate with the potassium permanganate.

POTASSIUM PERMANGANATE

Dilute 100 cm³ of the original oxalic acid to 250 cm³ and titrate 25 cm³ portions with permanganate.

(The smaller volume of permanganate solution needed in the first case is due to oxidation of some of the oxalic acid by the manganese (IV) oxide. Since both solutions were diluted in the same ratio of 100 : 250 the difference between the *total* volumes of potassium permanganate needed for titration measures the oxidising action of the MnO_2).

$$5\ MnO_2 \equiv 2\ KMnO_4$$
$$5(55 + 32)\ g \equiv 100\ dm^3\ 0{\cdot}02M.$$

Exercises

An asterisk indicates that the particular problem can be attacked without previous preparation of solutions, the concentrations of which have not to be divulged to the student.

(1)* [Solution of formic acid 1–2 g/dm³; 0·02M potassium permanganate; 0·1M sodium hydroxide.]

1. Titrate the solution of formic acid (HCO_2H) with the standard sodium hydroxide solution, and thereby determine its concentration in mole/dm³.
2. Add to 25 cm³ of the formic acid solution an excess of sodium carbonate solution, heat almost to boiling and titrate while hot with the standard potassium permanganate solution until the clear liquid above the precipitate is coloured pink.
3. Titrate 25 cm³ of the solution of formic acid, heated to boiling but without neutralisation with the sodium carbonate, with the potassium permanganate solution.
4. Interpret your results as far as you can.

(Oxford Schol.)

(2) [0·0167M potassium dichromate; 0·05M iron(III) sulphate; 0·05M titanium(II) sulphate; solution of hydroxylamine in sulphuric acid containing about 1·5 g/dm³.]

Hydroxylamine (NH_2OH) can be reduced to NH_3 or oxidised to definite stages such as N_2, N_2O, HNO_2 or HNO_3. You are given standard solutions of potassium dichromate, iron(III) sulphate and titanium (II) sulphate and a solution of hydroxylamine in sulphuric acid. Given that titanium(II) sulphate reduces hydroxylamine quantitatively to ammonia determine the nitrogen compound to which it is oxidised by (*a*) iron(III) sulphate, (*b*) potassium dichromate.

(Oxford Schol.)

Questions and Calculations

(1) Starting with manganese(IV) oxide, how would you prepare in the laboratory a crystalline specimen of potassium permanganate?
Why are 3·16 g/dm³ of potassium permanganate required to make a 0·02M solution?
State very briefly how a standard solution of permanganate can be used to determine the concentrations of solutions of

(a) an iron(II) salt;
(b) oxalic acid;
(c) hydrogen peroxide.

(2) Calculate the volume of a 0·02M solution of potassium permanganate required for the complete oxidation of 1 g of iron(II) oxalate (FeC_2O_4).

(L.)

(3) 50 cm³ of a solution of hydrogen peroxide were diluted to 1 dm³ with water, 25 cm³ of this solution, when acidified, reacted with 23·9 cm³ of 0·02M $KMnO_4$. Calculate the concentration of the original hydrogen peroxide solution in g/dm³.

(4) Calculate x in the formula, $FeSO_4,xH_2O$, from the following data: 12·20 g of iron(II) sulphate crystals were made up to 500 cm³ of acidified solution. 25 cm³ of this solution required 21·9 cm³ 0·02M $KMnO_4$.

(5) 1·500 g of iron were converted to 250 cm³ of acidified iron(II) sulphate solution, 25 cm³ of which were found to require 24·4 cm³ of 0·0218 $KMnO_4$. Calculate the percentage by weight of iron in the wire.

(6) Calculate the weights of anhydrous sodium oxalate and crystalline oxalic acid, $H_2C_2O_4,2H_2O$, per 1000 cm³ of a solution from the following data: 25 cm³ of the solution required 19·8 cm³ of 0·1M NaOH for neutralisation, using phenolphthalein as indicator; 25 cm³ of the solution required 34·2 cm³ of 0·0212M $KMnO_4$ for oxidation.

(7) Calculate the percentage of manganese(IV) oxide in a sample from the following data: 25 cm³ of a certain oxalic acid solution was titrated with 0·0214M $KMnO_4$. After prolonged boiling of the same volume of the oxalic acid solution with 0·142 g of the manganese(IV) oxide and dilute sulphuric acid, the volume of the same $KMnO_4$ solution required was reduced by 24·4 cm³.

(8) Calculate the weights of anhydrous iron(II) sulphate and iron(III) sulphate per 1000 cm³ of a solution given that 25 cm³ of it needed 21·6 cm³ of 0·021M $KMnO_4$ for oxidation while, after reduction of the same volume of the solution by zinc amalgam, 22·8 cm³ more of the $KMnO_4$ solution were needed for oxidation.

5. Potassium Dichromate

POTASSIUM dichromate is used as an oxidising agent in reactions very similar to those described in the last chapter in which potassium permanganate was employed, especially for titration of iron(II) salts.

It will be remembered that potassium permanganate is not used with chlorides because (especially in the presence of iron(III) salts) it may be used up in oxidising hydrochloric acid. Potassium dichromate is free from this disadvantage and may be used in the presence of hydrochloric acid or its salts. Until recently, the great disadvantage of potassium dichromate has been that one of its reduction products is a chromium(III) salt. This chromium salt is deep green in colour and effectively masks the colour of any excess of potassium dichromate, which cannot act (like potassium permanganate) as its own indicator.

Standard Potassium Dichromate Solution. In acid solution each "molecule" of potassium dichromate will react with six electrons

$$Cr_2O_7^{2-} + 8H^+ + 6e^- = 2Cr^{3+} + 4H_2O$$

or, in the molecular form with, for example, iron(II) chloride

$$K_2Cr_2O_7 + 14HCl + 6FeCl_2 = 2KCl + 2CrCl_3 + 6FeCl_3 + 7H_2O$$

Molecular weight 294.

Hence for 1 dm³ 0·0167M solution the weight of $K_2Cr_2O_7$ is $294 \times 0{\cdot}0167$ g = 4·90 g.

The solution may be made up directly from the pure salt if it is first melted in a porcelain dish, so that no water remains in it, and ground to powder after cooling.

Indicator. The indicator necessary is diphenylamine, and this compound, which is used in concentrated sulphuric acid solution, has no influence on the colour of the mixture, but, when oxidised by a slight excess of potassium dichromate, it produces an intensely coloured blue compound. It is necessary, however, to prevent iron(III) ions, which are formed in the course of titration of an iron(II) salt, from oxidising the diphenylamine prematurely, *i.e.*, before potassium dichromate is in excess, and for this purpose phosphoric acid is used. The iron(III) ions are then taken up, as fast as they are produced by oxidation, into an iron(III) phosphate complex which is almost undissociated and the iron(III) ions are prevented from oxidising the diphenylamine.

The titration is carried out in the following way: Make a solution of 1 g of diphenylamine in 50 cm³ of concentrated sulphuric acid.

50 THE ESSENTIALS OF VOLUMETRIC ANALYSIS

Dilute some syrupy phosphoric acid with twice its bulk of water. To 25 cm³ of the solution to be titrated add one drop of the diphenylamine solution and 5 cm³ of the diluted phosphoric acid. Titrate with potassium dichromate solution till the mixture (which is deep green in colour) just turns blue.

Potassium dichromate solution may be used for the estimation of iron and its salts by any of the methods previously described employing potassium permanganate. The following estimation is instructive. Potassium permanganate could not be used because of the presence of chlorides.

Experiment 27. Estimation of the percentage purity of a sample of metallic tin. (Tin; iron(III) chloride solution; 0·0167M potassium dichromate.)

Method. The tin is converted to tin(II) chloride solution, which is then used to reduce a portion of an iron(III) solution to the iron(II) state. The iron(II) solution is then oxidised back to the iron(III) condition by the potassium dichromate solution.

The reducing action of tin(II) chloride solution on an iron(III) salt may be expressed ionically in the form:

$$Sn^{2+} + 2Fe^{3+} = Sn^{4+} + 2Fe^{2+}$$

or as a partial equation

$$Sn^{2+} \rightarrow Sn^{4+} + 2e^-$$

From this for 250 cm³ of 0·05M $SnCl_2$ solution $119 \times 0·250 \times 0·05$ g or 1·49 g of tin are required.

Weigh out accurately about 1·5 g of tin, place in a conical flask and add about 50 cm³ of concentrated hydrochloric acid diluted with its own volume of water. Put a funnel into the neck of the flask to prevent loss by splashing and when the tin is completely dissolved transfer the tin(II) chloride solution to a 250 cm³ flask. Wash the funnel and beaker well with acid of the concentration previously used—excess hydrochloric acid is necessary to prevent precipitation of basic tin salts—and add the washings to the main solution, finally making up to the mark with distilled water.

Make up a solution containing about 9 g of iron(III) chloride in 250 cm³ dilute hydrochloric acid. Measure out 25 cm³ of it by a pipette into a conical flask and add 25·0 cm³ of the tin(II) chloride solution. Since the iron(III) solution has the higher concentration all the tin(II) ions are converted into tin(IV) ions. Titrate the mixture with potassium dichromate solution, using indicator and conditions as described on p. 49. Repeat to obtain concordant results.

Calculation. Suppose a g of tin were converted into 250 cm³ of tin(II) chloride solution and 25 cm³ of this reduced the iron(III) chloride solution so that b cm³ of 0·0167M $K_2Cr_2O_7$ were required for its oxidation.

POTASSIUM DICHROMATE

$$Sn + 2HCl = SnCl_2 + H_2$$
$$SnCl_2 + 2FeCl_3 = SnCl_4 + 2FeCl_2$$
$$6FeCl_2 + K_2Cr_2O_7 + 14HCl = 6FeCl_3 + 2KCl + 2CrCl_3 + 7H_2O$$

From these equations calculate the molarity and hence the weight of pure tin taken, then the percentage purity of the tin

$$= \frac{(3 \times 119) \times 10b \times 100}{60\,000 \times a}.$$

Experiment 28. To determine the percentage purity of a sample of potassium chromate. (Potassium chromate; 0·1M iron(II) ammonium sulphate.)

Potassium chromate is converted by acids into potassium dichromate.

$$2CrO_4^{2-} + 2H^+ = Cr_2O_7^{2-} + H_2O$$
or $\qquad 2K_2CrO_4 + H_2SO_4 = K_2SO_4 + K_2Cr_2O_7 + H_2O.$

Calculate the weight of potassium chromate needed to produce 250 cm³ 0·0167M $K_2Cr_2O_7$ when in acidified solution (p. 49). Weigh out this amount of potassium chromate (by difference) and including in the solution 50 cm³ of bench sulphuric acid, make up to 250 cm³ with distilled water.

Titrate this, from a burette, against the iron(II) ammonium sulphate solution (p. 49), using diphenylamine as indicator. Calculate the percentage purity from the equations given above.

Experiment 29. To determine the percentage of potassium chlorate in a given sample. (Potassium chlorate; 0·2M iron(II) ammonium sulphate; 0·0167M potassium dichromate.)

In hot acid solution, potassium chlorate[1] oxidises iron(II) salts.

$$6Fe^{2+} + 6H^+ + ClO_3^- = 6Fe^{3+} + Cl^- + 3H_2O$$
i.e., $\qquad KClO_3 \equiv 6 \text{ dm}^3 \ 0·0167M \ K_2Cr_2O_7.$

Calculate the weight of potassium chlorate which is required for 250 cm³ 0·1M solution and weigh out this amount. Make it up to 250 cm³ of solution with distilled water.

Prepare 250 cm³ of 0·2M iron(II) ammonium sulphate solution (p. 38). To 25 cm³ of this solution, add 25 cm³ of the potassium chlorate solution, about 15 cm³ of dilute sulphuric acid, and boil the mixture, with a funnel in the mouth of the flask, for about twenty minutes. Repeat this twice to obtain three results.

Then wash the funnels, cool the mixture and titrate with the dichromate solution. Calculate the percentage purity from the equations given above.

[1] The percentage purity of manganese(IV) oxide, trilead tetroxide and many other oxidising agents can be obtained by using these substances in the place of potassium chlorate in Expt. 29.

52 THE ESSENTIALS OF VOLUMETRIC ANALYSIS

Experiment 30. To determine the percentage of iron in a sample of iron alum, reducing the iron by tin(II) chloride. (Iron alum; 0·0167M potassium dichromate; tin(II) chloride.)

Prepare a solution of iron alum of accurately known concentration and roughly 0·1M.

Prepare a solution of tin(II) chloride by dissolving about 3 g of the salt ($SnCl_2,2H_2O$) in 50 cm^3 of concentrated hydrochloric acid and diluting to 250 cm^3.

Add 5 cm^3 of concentrated hydrochloric acid to 25 cm^3 of the alum solution in a conical flask and boil the liquid. Run in the tin(II) chloride solution drop by drop from a burette until the yellow colour of the iron(III) ions is discharged. Add two or three more drops of the tin(II) chloride solution to make certain that the reduction is completed. (Care should be taken not to add too great an excess of tin(II) chloride.) Now add 1 or 2 cm^3 of a saturated solution of mercury(II) chloride. A small white precipitate should be observed showing that sufficient tin(II) chloride had been added. Titrate this solution with the dichromate using diphenylamine as indicator.

Experiment 31. Determine the percentage of iron in a sample of spathic iron ore (iron(II) carbonate). (Iron ore, zinc amalgam. 0·0167M potassium dichromate.)

Most of the iron in this ore is in the iron(II) but some is in the iron(III) condition.

The molecular weight of iron(II) carbonate is 116 so to obtain 250 cm^3 of an approximately 0·1M iron(II) solution, weigh out about $116 \times 0·250 \times 0·1$ g of the ore (*i.e.*, about 3 g).

Weighing by difference, put the ore into a conical flask and add about 50 cm^3 of bench hydrochloric acid. Cover the mouth of the flask with a small watch-glass and warming the mixture and adding more acid if necessary, allow the action to continue till no brown particles remain. (Silica will remain undissolved.)

Wash the watch-glass and reduce all the iron to the iron(II) state by zinc amalgam (p. 42). Make up the reduced solution to 250 cm^3 in a measuring flask. Titrate with dichromate solution, using diphenylamine as indicator. Calculate the percentage of iron in the ore from the above equations.

Exercises

An asterisk indicates that the particular problem can be attacked without previous preparation of solutions, the concentrations of which have not to be divulged to the student.

(1)* Oxidising power of $KMnO_4$ and $K_2Cr_2O_7$. [Tenth molar solution of each. Iron(II) ammonium sulphate, diphenylamine.]

You are supplied with tenth molar solutions of potassium permanganate

POTASSIUM DICHROMATE

and potassium dichromate and iron(II) ammonium sulphate. Compare the oxidising powers of the two solutions in acid solutions by oxidation of a solution of iron(II) ammonium sulphate (this need not be made up accurately). Explain your result by means of equations.

(2) Estimation of potassium sulphate and potassium dichromate in a mixture of the two. [Mixture containing potassium dichromate and potassium sulphate. Iron(II) ammonium sulphate.]

Make a solution of the mixture to be approximately 0·0167M with respect to the dichromate (assume the solid to contain 50% of the dichromate). Make up a standard solution of iron(II) ammonium sulphate and titrate with the unknown solution using diphenylamine as an indicator.

Calculations

(1) 2 g of a specimen of hydrated iron(III) oxide were dissolved in hydrochloric acid and the solution was made up to 250 cm³. 25 cm³ of this solution after reduction to the iron(II) state required 13·7 cm³ 0·0167M $K_2Cr_2O_7$ for oxidation. Calculate the percentage of Fe_2O_3 in the specimen.

(2) 8·3 g of pure tin were dissolved in hydrochloric acid and the solution was made up to 1 dm³. 25 cm³ of this solution required 26·7 cm³ of a solution of potassium dichromate for complete oxidation. Calculate the concentration of the dichromate solution in g/dm³.

6. Iodine and Sodium Thiosulphate

SODIUM thiosulphate reacts with iodine, producing sodium iodide and sodium tetrathionate

$$I_2 + 2Na_2S_2O_3 = Na_2S_4O_6 + 2NaI$$

or ionically $\quad I_2 + 2S_2O_3^{2-} = 2I^- + S_4O_6^{2-}$

This reaction may be used for the estimation of iodine or, indirectly, for the estimation of a substance which participates in a reaction in which iodine is liberated.

Standard Sodium Thiosulphate Solution. The sodium thiosulphate purchased for laboratory purposes is the hydrated salt $Na_2S_2O_3,5H_2O$ (molecular weight 248).

From the above equation:—$2S_2O_3^{2-} = S_4O_6^{2-} + 2e^-$. From this 1 dm³ of 0·1M sodium thiosulphate solution contains 24·8 g of the pure hydrated salt. An accurately standard solution of the salt cannot be made up directly from the salt as usually purchased because it is not sufficiently pure. Make up a roughly 0·1M solution by dissolving 25 g of sodium thiosulphate crystals in warm distilled water and diluting to 1 dm³ in a measuring flask. This solution must now be standardised. This is usually carried out by the use of either accurately 0·02M potassium permanganate or pure potassium iodate (KIO_3).

Indicator and End-point. The indicator used in titrating iodine solutions with sodium thiosulphate is starch solution. With free iodine it produces a deep blue coloration, the blue colour disappearing as soon as sufficient sodium thiosulphate has been added to react with all the iodine.

The procedure is as follows: Sodium thiosulphate solution is run from a burette into the iodine solution until its original brown colour is changed to pale yellow. Then a few drops of starch solution are added producing a blue coloration. Further drop-by-drop addition of thiosulphate solution is continued until the blue coloration disappears.

Preparation of the Starch Indicator. Mix 1 g of starch to a thin paste with water in an evaporating dish, then pour the paste into about 350 cm³ of boiling water. Boil the mixture for two to three minutes then cool it for use. (This solution will not keep; moulds will grow on it. The addition of about 0·5 g of salicylic acid to the 250 cm³ of water before boiling will prevent such growth and the starch indicator will then keep for a long time.)

IODINE AND SODIUM THIOSULPHATE 55

Experiment 32. Standardisation of sodium thiosulphate solution by potassium permanganate. (0·02M potassium permanganate; approx. 0·1M sodium thiosulphate; potassium iodide.)

The permanganate is added to acidified potassium iodide solution. Iodine is liberated and is titrated with the thiosulphate solution.

$$2KMnO_4 + 10KI + 16HCl = 12KCl + 2MnCl_2 + 8H_2O + 5I_2$$

ionically, $2MnO_4^- + 10I^- + 16H^+ = 2Mn^{2+} + 8H_2O + 5I_2$.

Dissolve about 1 g[1] of potassium iodide in about 20 cm³ distilled water in a conical flask and acidify with an equal volume of 3M hydrochloric acid.

Add 25 cm³ 0·02M potassium permanganate and dilute the brown solution to about 100 cm³. From a burette, add the sodium thiosulphate solution until the mixture is pale yellow; add the starch indicator and continue titration till the blue colour is discharged. Repeat the titration two or three times.

Calculation. Suppose 25 cm³ 0·02M $KMnO_4$ liberate iodine requiring 24·7 cm³ of the sodium thiosulphate solution.

Then the sodium thiosulphate solution is $\frac{25}{24\cdot7} \times 0\cdot02 \times 5M$ or 0·1012M. The concentration of the hydrated salt is $248 \times 0\cdot1012 = 25\cdot1$ g/dm³.

0·0167M potassium dichromate solution may be substituted for 0·02M permanganate, the estimation being otherwise identical.

Experiment 33. Standardisation of sodium thiosulphate solution by pure potassium iodate. (Approximately 0·1M sodium thiosulphate; potassium iodate; potassium iodide.)

Standard potassium iodate solution reacts with excess potassium iodide in acidified solution and the liberated iodine is titrated with the sodium thiosulphate solution,

ionically $IO_3^- + 5I^- + 6H^+ = 3H_2O + 3I_2$

or $KIO_3 + 5KI + 3H_2SO_4 = 3K_2SO_4 + 3H_2O + 3I_2$

Molecular weight 214.

For 250 cm³ 0·0167 M.KIO_3 solution $214 \times 0\cdot25 \times 0\cdot0167$ g or 0·892 g are needed.

Make up a solution of potassium iodate to contain 0·892 g in 250 cm³. Pipette 25 cm³ of it into a conical flask and add about 1 g of potassium iodide. Acidify the solution with about 10 cm³ of bench sulphuric acid and titrate the liberated iodine with the sodium

---

[1] $2I^- \rightarrow I_2 + 2e^-$
Molecular weight of KI = 166
For 25 cm³ 0·1M solution, $166 \times \frac{25}{10\,000}$ g of potassium iodide, *i.e.*, 0·4 g are needed. 1 g potassium iodide provides an excess for solution of the liberated iodine—alternatively use about 10 cm³ of a 10% solution of potassium iodide in each titration.

56 THE ESSENTIALS OF VOLUMETRIC ANALYSIS

thiosulphate solution as described before. Repeat the titration two or three times. The calculation is similar to that given in Experiment 32.

Experiment 34. To prepare a standard iodine solution. (Iodine; potassium iodide; 0·1M sodium thiosulphate.)

Iodine is only very slightly soluble in water. It is, however, readily soluble in potassium iodide solution forming a brown liquid which contains the compound KI_3. This compound liberates iodine so readily that the solution behaves as if the dissolved iodine were all free iodine.

$$I_2 + I^- \rightleftharpoons I_3^-$$

Iodine can be purchased sufficiently pure to justify direct preparation of a standard solution, but it is volatile enough to make it almost impossible to avoid loss while the solution is being prepared. Hence, standardisation is necessary.

{ 0.01M 1 dm³ of 0·05M iodine solution contains 12·7 g. Weigh out 13 g of iodine in a weighing bottle. Transfer it to a 1 dm³ flask. Dissolve about 25 g of potassium iodide in 100 cm³ of water and add about 80 cm³ of it to the iodine, using the remaining 20 cm³ for washing out the weighing bottle. When the iodine is dissolved, dilute the solution to 1 dm³ and shake well.

Titrate 25 cm³ of this solution with standard 0·1M sodium thiosulphate solution. The calculation is similar to that given on p. 55.

Experiment 35. Estimation of available chlorine in bleaching powder. (Bleaching powder; potassium iodide; acetic acid; 0·1M sodium thiosulphate.)

Note. Bleaching powder liberates chlorine when reacting with a dilute acid; this chlorine is available for bleaching and is known as "available" chlorine.

Bleaching powder may deteriorate for two reasons:—

(i) Because it is attacked by carbon dioxide of the air
(ii) Because of internal changes.

Both these changes reduce the "available" chlorine content and an old sample of bleaching powder may be almost worthless for bleaching. The following method estimates "available" chlorine. Solutions of household bleaches can be estimated along similar lines.

Method. Bleaching powder is mixed with potassium iodide solution and the mixture is acidified. The liberated iodine is titrated by sodium thiosulphate solution. Acetic acid is used to liberate the chlorine being quicker than atmospheric carbon dioxide.

A mineral acid, such as hydrochloric, would allow calcium chlorate in the bleaching powder to liberate iodine from the potassium iodide, giving a result which is too high.

IODINE AND SODIUM THIOSULPHATE

$$-Cl_2 + 2KI = 2KCl + I_2$$
$$\text{or } Cl_2 + 2I^- \rightarrow 2Cl^- + I_2$$

To allow for impurity in the bleaching powder and for deterioration in "available" chlorine content, about 2·5 g should be used. Weigh accurately a weighing bottle containing about 2·5 g of bleaching powder which has been powdered as finely as possible. Prepare in a 250 cm³ measuring flask a solution of 4 to 5 g of potassium iodide in 20 cm³ of water. Transfer the bleaching powder to a clean mortar and weigh the bottle again. Rub the bleaching powder into a paste with a little water at a time, transferring the paste to the 250 cm³ flask, until all the bleaching powder is in the flask. Acidify the mixture with acetic acid and make up the clear brown solution to 250 cm³. Shake the flask well and titrate 25 cm³ of the solution against the sodium thiosulphate solution as described previously, using starch indicator. Repeat the titration two or three times.

Calculation. Suppose a g of bleaching powder were used and 25 cm³ of the resulting iodine solution required b cm³ 0·1M $Na_2S_2O_3$.

$$Cl_2 \equiv I_2 \equiv 2Na_2S_2O_3$$
$$(2 \times 35 \cdot 5) \text{ g} \qquad 2 \text{ dm}^3 \text{ M}$$

Weight of available chlorine $= (2 \times 35 \cdot 5) \times \dfrac{b}{20\,000}$ g

Weight of available chlorine per 250 cm³ of solution

$$= (2 \times 35 \cdot 5) \times \dfrac{10b}{20\,000} \text{ g}$$

Percentage of available chlorine in the bleaching powder

$$= (2 \times 35 \cdot 5) \times \dfrac{10b}{20\,000} \times \dfrac{100}{a}$$

Experiment 36. Estimation of copper, e.g., determination of the percentage of copper sulphate in a sample of copper sulphate crystals. (Copper(II) sulphate; potassium iodide; acetic acid; 0·1M sodium thiosulphate.)

The copper solution must be free from anything but a trace of mineral acid. Otherwise the end-point is not accurate. This solution then liberates iodine from potassium iodide solution in accordance with the equation:—

$$2Cu^{2+} + 4I^- = 2CuI \downarrow + I_2$$
$$\text{or } Cu^{2+} + e^- \rightarrow Cu^+ \qquad \text{Copper (I) iodide}$$

The iodine is titrated by the sodium thiosulphate. From the equation

$$2Cu \equiv I_2 \equiv 2Na_2S_2O_3,$$
i.e., $\qquad CuSO_4,5H_2O \equiv Na_2S_2O_3.$

Molecular weight 249·5.

To produce 250 cm³ of 0·1M solution, 249·5 × 0·25 × 0·1 g of copper sulphate crystals are needed. Weigh accurately a weighing bottle containing about 6 g of copper sulphate crystals. Transfer the crystals to a 250 cm³ flask and weigh the weighing bottle again. Add sodium carbonate solution till a *slight* permanent bluish precipitate of copper carbonate has formed. Acidify the mixture with a little acetic acid when a clear blue solution is obtained. Make the solution up to 250 cm³ with distilled water and shake well. Pipette 25 cm³ of the solution into a conical flask containing about 1·5 g potassium iodide dissolved in a little water. Titrate the liberated iodine against the sodium thiosulphate solution as previously described. Calculate the percentage of copper sulphate present.

Experiment 37. Estimation of a sulphite, e.g., determination of the percentage of anhydrous sodium sulphite in a sample of the crystals. (0·05M iodine; 0·1M sodium thiosulphate; potassium iodide; sodium sulphite crystals.)

Standard sodium sulphite solution is added to an excess of standard iodine solution and the excess of iodine is estimated by sodium thiosulphate.

In the presence of sodium hydrogen carbonate, sodium sulphite is oxidised quantitatively by iodine to sodium sulphate.

$$Na_2SO_3 + I_2 + H_2O = Na_2SO_4 + 2HI$$
$$2HI + 2NaHCO_3 = 2NaI + 2H_2O + 2CO_2$$

From these

$$SO_3^{2-} + H_2O \rightarrow SO_4^{2-} + 2H^+ + 2e^-$$

Molecular weight of $Na_2SO_3, 7H_2O = 252$

i.e., for 250 cm³ 0·05M solution use 252 × 0·25 × 0·05 g or 3·15 g pure sodium sulphite crystals. Make up a solution containing 3·3 to 3·4 g of the impure crystals.

Pipette 50 cm³ of 0·05M iodine solution into a conical flask and from a pipette run into it, slowly and with constant shaking, 25 cm³ of the sodium sulphite solution. Add about 2 g of sodium hydrogen carbonate, shake the flask and titrate the excess iodine with the sodium thiosulphate solution, using starch indicator as previously described. Repeat the estimation two or three times.

Calculation. Suppose a g of sodium sulphite crystals were made up to 250 cm³ of solution and, after 25 cm³ of this solution had been added to 50 cm³ 0·05M iodine solution, b cm³ 0·1M thiosulphate solution were needed to titrate excess iodine, *i.e.*, b cm³ 0·05M iodine solution were left in excess.

$$Na_2SO_3 \equiv I_2$$
$$126 \text{ g} \quad 20 \text{ dm}^3 \text{ 0·05M}$$

25 cm³ of the sodium sulphite solution reacted with $(50 - b)$ cm³

IODINE AND SODIUM THIOSULPHATE

0·05M iodine solution. To oxidise the whole of the sodium sulphite, $10(50 - b)$ cm³ iodine solution would be needed.

∴ Weight of anhydrous sodium sulphite present
$$= \frac{126 \times 10 (50 - b)}{20\,000} \text{ g}$$

∴ Percentage of anhydrous sodium sulphite in the crystals
$$= \frac{126 \times 10(50 - b)}{20\,000} \times \frac{100}{a}$$

Exercises

An asterisk indicates that the particular problem can be attacked without previous preparation of solutions, the concentrations of which have not to be divulged to the student.

(1)* [Malachite; 0·1M sodium thiosulphate.] The substance A is an impure specimen of copper carbonate. Estimate by the following method the percentage weight of copper which the substance contains.

Weigh out accurately about 3 g of the specimen A, dissolve it with caution in the minimum amount of dilute hydrochloric acid, add sodium carbonate solution drop by drop until a faint precipitate appears and then add dilute acetic acid until this precipitate just dissolves. Make up the solution to 250 cm³ with distilled water. To 25 cm³ portions of this solution add a few crystals (not less than 1 g) of potassium iodide, and when these have dissolved titrate the liberated iodine with the sodium thiosulphate solution provided.

$$2CuCl_2 + 4KI = 2CuI \downarrow + 4KCl + I_2.$$

(J.)

(2) [Unknown potassium sulphate solution; barium chromate in dilute hydrochloric acid, about 5 g/dm³ suitable; potassium iodide; 0·1M sodium thiosulphate.]

Estimate the weight of potassium sulphate (K_2SO_4) per dm³ of the given solution A by the following method:

You are provided with a solution of barium chromate in dilute hydrochloric acid. To 100 cm³ of this solution at the boiling temperature add 20 cm³ of the sulphate solution A. Subsequently, cool and then add ammonia cautiously to the liquid until the colour changes from orange to yellow. Filter, and treat the filtrate with hydrochloric acid and potassium iodide. Titrate the iodine produced with the standard solution of sodium thiosulphate provided.

(Cambridge Schol.)

(3) [Unknown mixture of copper sulphate and sodium chloride, about 35 g/dm³; 0·051M iodine; sodium thiosulphate approximately 0·1M; potassium iodide 10% solution w/v.]

By means of the given solution of iodine determine the molarity of the solution of sodium thiosulphate and use this solution to determine the amount of copper sulphate ($CuSO_4$) in the mixture which contains copper sulphate and sodium chloride.

(L.)

60 THE ESSENTIALS OF VOLUMETRIC ANALYSIS

Questions and Calculations

(1) 0·200 g of the mineral pyrolusite was heated with excess of concentrated hydrochloric acid, and the gas evolved passed into excess of potassium iodide solution. The iodine released required 35·0 cm^3 0·1M solution of sodium thiosulphate for complete reaction. Calculate the percentage of manganese(IV) oxide in the pyrolusite.

(2) 500 cm^3 of a mixture of oxygen and ozone at s.t.p. are allowed to react with an acidified solution of potassium iodide. The iodine liberated reacts with 37·6 cm^3 0·1M sodium thiosulphate solution. Calculate the percentage by volume of ozone in the gas.
(L.)

(3) You are provided with a pure sample of potassium tetroxalate (KH$_3$(C$_2$O$_4$)$_2$,2H$_2$O; mol. wt. = 254·2) as reference substance and solutions of potassium permanganate and sodium thiosulphate. Explain how you would prepare and standardise a solution of iodine of the order of 0·05M. An accuracy of 0·1 % is required.

Sulphur dioxide reacts with iodine and water as follows:
$$SO_2 + I_2 + 2H_2O = H_2SO_4 + 2HI$$
10 cm^3 of a saturated solution of sulphur dioxide in water at 20°C are diluted to 1000 cm^3 with air free water. 25·0 cm^3 of this diluted solution are treated with 25·0 cm^3 0·05M solution of iodine (excess) and back-titrated with 0·1M of thiosulphate, requiring 16·70 cm^3. Calculate the concentration of the saturated solution of sulphur dioxide in g/dm^3.

Why is this method of back-titration employed?

(4) 1 g of pure potassium permanganate and 1 g of pure potassium dichromate were dissolved in water and excess acidified potassium iodide solution was added. What volume of double molar sodium thiosulphate would be required to react with the iodine so produced?

(5) Give an account of the chemistry underlying the use of iodine and potassium iodide in volumetric analysis. Illustrate your answer by reference to *three* examples in addition to the reactions mentioned below.

25 cm^3 of a solution of potassium permanganate on mixing with excess of potassium iodide in the presence of dilute sulphuric acid liberated iodine which reacted with exactly 40 cm^3 of decimolar sodium thiosulphate.

What was the concentration of the potassium permanganate solution in g/dm^3?

(6) A neutral solution of potassium permanganate is said to react with a hot solution of potassium iodide according to the equation—
$$2KMnO_4 + KI + H_2O = 2MnO_2 + KIO_3 + 2KOH$$
Describe the experiments which you would undertake in order to verify this statement.
(O.)

7. Silver Nitrate

SILVER nitrate is a non-hygroscopic solid capable of being prepared in a high degree of purity, and hence very satisfactory for use in volumetric analysis. Its use depends upon the insolubility of its halogen salts and it is commonly used to estimate chlorides.

$$AgNO_3 + NaCl = NaNO_3 + AgCl$$

or ionically $\quad Ag^+ + Cl^- = AgCl$

It can be used as its own indicator, for example, by running a solution of the nitrate into a solution of common salt until on allowing the precipitate of silver chloride to settle, a further drop produces no precipitate. This is the end-point. It is clear that this method would be tedious and difficult for a beginner. Hence it is usual to employ as indicator potassium chromate solution. If silver nitrate solution is added to a sodium chloride solution containing a few drops of potassium chromate solution the silver chloride is selectively precipitated before the silver chromate. One drop of the silver nitrate solution in excess will produce sufficient silver ions to exceed the solubility product of the silver chromate, and hence that is precipitated as a brick-red substance, and the end-point is the first appearance in the mixture of a reddish tinge

$$2Ag^+ + CrO_4^{2-} \rightleftharpoons Ag_2CrO_4$$

Limitations of Use of Potassium Chromate. Potassium chromate can only be used in neutral solutions because

(a) Silver chromate is soluble in acids;
(b) An alkaline solution would react with the silver nitrate to form silver oxide.

$$2Ag^+ + 2OH^- = Ag_2O \downarrow + H_2O$$

If an acidic solution of a chloride is to be estimated there are available the following alternatives:—

(a) Neutralise the acid with excess of calcium carbonate (free from chloride) and use potassium chromate as an indicator. (See p. 62.)
(b) Use an adsorption indicator.
(c) Add excess silver nitrate and estimate the excess by potassium thiocyanate. (See p. 70.)

These titrations are very accurate and therefore solutions of silver nitrate are usually used in concentrations approximately tenth molar or less.

N.B. Solutions of silver nitrate must be made up with distilled water because tap water usually contains dissolved chloride.

Experiment 38. Standardisation of silver nitrate solution. (Silver nitrate; pure sodium chloride.)

A standard solution of silver nitrate can be made up by weighing the silver nitrate directly. It may be necessary, however, to standardise a solution of silver nitrate of unknown concentration.

$AgNO_3 + HCl = AgCl + HNO_3$
Molecular weight 170.

∴ 250 cm³ 0·1M $AgNO_3$ contains $170 \times 0.25 \times 0.1 = 4.25$ g.

Weigh accurately a weighing bottle containing about 4·25 g of small crystals of silver nitrate. Transfer these to a 250 cm³ flask and weigh the bottle accurately. Add distilled water to the flask and shake after each addition. When all the silver nitrate has dissolved, add distilled water up to the mark by means of a pipette and shake well.

$NaCl + AgNO_3 = AgCl + NaNO_3$
Molecular weight 58·5.

∴ 250 cm³ 0·1M NaCl solution should contain 1·46 g.

Proceed in exactly the same way as the above to make an approximately 0·1M solution of sodium chloride, weighing out accurately about 1·5 g of pure dry sodium chloride.

Take 25 cm³ of the sodium chloride solution, add 1 cm³ of a 5% solution of potassium chromate and run in the silver nitrate from the burette. Proceed slowly towards the end-point and take the reading when the first permanent reddish tinge is apparent. An allowance may be made for the nitrate used to affect the indicator by performing a "blank" experiment, using 1 cm³ potassium chromate solution in 50 cm³ of water. Repeat the experiment two or three times until concordant results are obtained.

Calculation. Suppose weight of NaCl taken = 1·52 g in 250 cm³.
Suppose 25 cm³ of NaCl solution required 26·2 cm³ of the silver nitrate solution.

Concentration of NaCl = 1.52×4 g/dm³ = 6·08 g/dm³.

∴ Molarity of NaCl solution is $\dfrac{6.08}{5.85} \times 0.1$

∴ Molarity of $AgNO_3$ solution is

$$\frac{25}{26.2} \times \frac{6.08}{5.85} \times 0.1 = 0.0992$$

Concentration of $AgNO_3 = \dfrac{25}{26.2} \times \dfrac{6.08}{5.85} \times 0.1 \times 170$ g/dm³
= 16·9 g/dm³

Experiment 39. Standardisation of hydrochloric acid by means of silver nitrate. (Approx. 0·1M hydrochloric acid; 0·1M silver nitrate; Chloride-free calcium carbonate.]

You will probably have prepared a solution of hydrochloric acid approximately 0·1M. This may be standardised by means of the silver nitrate solution.

Use of Calcium Carbonate. This will react with the acid (any unused carbonate remaining undissolved) producing an equivalent amount of chloride in neutral solution. The carbonate removes the free hydrogen ions which would affect the indicator and leaves the chloride ions undisturbed.

$$CaCO_3 + 2HCl = CaCl_2 + H_2O + CO_2$$
or $$CO_3^{2-} + 2H^+ = + H_2O + CO_2 \uparrow$$

From this equation
100 g of $CaCO_3$ would neutralise 2 dm³ M.HCl
25 cm³ 0·1M HCl require 0·125 g of $CaCO_3$
This need not be weighed out—use a liberal excess—say about 1 g.

Method. Run 25 cm³ of the hydrochloric acid from a pipette into a conical flask and add about 1 g of precipitated chalk. Add 1 cm³ of 5% potassium chromate solution and run in the silver nitrate from the burette until the first permanent reddish tinge is observed. Repeat twice or until two concordant results are obtained.

$$CaCl_2 + 2AgNO_3 = 2AgCl + Ca(NO_3)_2$$
$$\therefore 2HCl \equiv 2AgNO_3$$

Calculation. Suppose 25 cm³ of the acid required 24·3 cm³ of 0·0993 M.$AgNO_3$.

$\therefore$ Molarity of HCl solution $= \dfrac{24 \cdot 3}{25} \times 0 \cdot 0993$

$\therefore$ Concentration of HCl $= 0·0965$
solution $\qquad\qquad = 0·0965 \times 36·5$ g/dm³
$\qquad\qquad\qquad = 3·52$ g/dm³

Experiment 40. Determination of amounts of sodium and potassium chloride in a mixture. [Mixture[1] of potassium and sodium chlorides (about 3 g of each to make 1 dm³ of solution) 0·1M silver nitrate.]

$$KCl + AgNO_3 = AgCl + KNO_3$$
Molecular weight 74·5
$$NaCl + AgNO_3 = AgCl + NaNO_3$$
Molecular weight 58·5.
In both cases $\qquad Ag^+ + Cl^- \rightarrow AgCl \downarrow$

Assuming the mixture to consist of approximately equal proportions of sodium and potassium chloride 1 dm³ M solution would contain about 65 g. Hence amount to be dissolved in 250 cm³ to make solution 0·1 M $= 65 \times 0·250 \times 0·1$ g $= 1·6$ g (approx.).

Weigh accurately a weighing bottle containing about 1·6 g of the

[1] After making up the mixture it is advisable to heat it gently for some time and allow to cool in a desiccator.

64 THE ESSENTIALS OF VOLUMETRIC ANALYSIS

mixture. Transfer the solid to a 250 cm³ flask and again weigh the bottle. Make up the solution to 250 cm³ in the distilled water and shake well. Titrate 25 cm³ with 0·1M $AgNO_3$ using 1 cm³ of 5% potassium chromate solution as an indicator.

Take the burette reading at the first permanent reddish tinge and repeat the experiment two or three times.

Calculation. Suppose weight of mixed chlorides = 1·550 g
Suppose 25 cm³ of solution required 23·1 cm³ 0·1M $AgNO_3$.
Let the 1·55 g of mixture contain x g of sodium chloride.
From the equation above
58·5 g of NaCl would require 10 000 cm³ 0·1M $AgNO_3$.

$$\therefore x \text{ g} \quad ,, \quad ,, \quad ,, \quad \frac{x}{58 \cdot 5} \times 10\ 000 \text{ cm}^3$$

Similarly (1·55 − x) g of KCl would require
$$\frac{(1 \cdot 55 - x)}{74 \cdot 5} \times 10\ 000 \text{ cm}^3 \ 0 \cdot 1\text{M AgNO}_3.$$

But 25 cm³ of the solution required 23·1 cm³
∴ 250 cm³ of the solution would require 231 cm³

Hence $\dfrac{x}{58 \cdot 5} \times 10\ 000 + \dfrac{(1 \cdot 55 - x)}{74 \cdot 5} \times 10\ 000 = 231$.

Then $x = 0 \cdot 623$ so weight of NaCl = 0·623 g and
$(1 \cdot 55 - x) = 0 \cdot 927$ g = weight of KCl.
∴ % of NaCl by weight = 40.   % of KCl by weight = 60.

N.B. The results from this experiment cannot be expected to show the same accuracy as an ordinary estimation, for example, of the amount of sodium chloride in a solution. An error of 0·1 cm³ in a titration requiring about 25 cm³ $AgNO_3$ would produce an error of about 5% in the above estimation of the amount of sodium chloride in a mixture, but less than 0·5% in the estimation of the amount of sodium chloride in a solution.

Experiment 41. To determine the number of molecules of water of crystallization in barium chloride crystals. (0·1M silver nitrate; barium chloride crystals; sodium sulphate.)

If this experiment is to be performed using potassium chromate solution as indicator then the barium ions must first be removed from the solution, since the reaction
$$Ba^{2+} + CrO_4^{2-} = BaCrO_4 \downarrow$$
would remove the indicator from the solution. The barium ions can be removed by adding excess sodium sulphate, which removes the barium as insoluble barium sulphate.
$$Ba^{2+} + SO_4^{2-} = BaSO_4 \downarrow$$
The result is, in effect, that an equivalent solution of sodium chloride is titrated instead of the barium chloride solution.

SILVER NITRATE

Quantities. $BaCl_2 + Na_2SO_4 = BaSO_4 + 2NaCl$
Molecular weights 208.   142.
$$NaCl + AgNO_3 = AgCl + NaNO_3$$
From these equations it is evident that

208 g $BaCl_2$ requires $\begin{cases} 142 \text{ g } Na_2SO_4 \text{ to precipitate the barium.} \\ 2 \text{ dm}^3 \text{ M.AgNO}_3 \text{ to precipitate the chloride.} \end{cases}$

Hence weight of *anhydrous* barium chloride in 250 cm³ of 0·05M solution
$$= 208 \times 0.250 \times 0.05 \text{ g} = 2.60 \text{ g}$$
and 25 cm³ of this solution would require
$$142 \times \frac{25}{20\,000} \text{ g} = 0.18 \text{ g of sodium sulphate.}$$

(Hence about 1 g of the anhydrous solid (or 2 g of the crystals $Na_2SO_4,10H_2O$) would provide ample excess.)

Allowing for the presence of water of crystallization a suitable weight of barium chloride to use would be about 3 g.

Method. Weigh out accurately (by means of a weighing bottle) about 3 g of barium chloride crystals and make up to 250 cm³ with distilled water. Shake well. Take 25 cm³ of this solution by means of a pipette, transfer to a conical flask, add about 1 g of anhydrous sodium sulphate and shake. Add about 1 cm³ of 5% potassium chromate solution and run in silver nitrate from a burette until a permanent reddish tinge is observed.

Repeat the experiment two or three times.

Calculation. Suppose weight of barium chloride crystals made up to 250 cm³ is a g and 25 cm³ of this solution required b cm³ of silver nitrate solution of concentration c molar.

Calculate the molarity of the barium chloride and hence the weight of solute in 250 cm³ solution.

Hence $\dfrac{BaCl_2}{BaCl_2, xH_2O} = \dfrac{b.c}{25} \times \dfrac{208}{8a}$

$\therefore 208 \times a = (208 + 18x)\dfrac{b.c}{25} \times \dfrac{208}{8}$ from which x can be obtained.

Experiment 42. Estimation of chloride and alkali in a solution containing them both. (0·1M hydrochloric acid; 0·1M silver nitrate. Alkaline sodium chloride solution.) A solution of suitable concentration can be quickly made by putting a known weight (about 3 g) of sodium chloride in water in a 1000 cm³ flask. Add 500 cm³ of 0·1M NaOH solution and make up to the mark.

Principle. Silver nitrate cannot be used in alkaline solution with potassium chromate as an indicator. Hence the alkali is neutralised with standard hydrochloric acid and the total chloride present is estimated with standard silver nitrate solution.

Equations. $NaOH + HCl = NaCl + H_2O$
$NaCl + AgNO_3 = NaNO_3 + AgCl$

Method. Take 25 cm³ of the alkaline sodium chloride solution and add a few drops of phenolphthalein solution. It will turn red. Run hydrochloric acid into this solution until the pink colour is just discharged. Note the reading. Add 1 cm³ of 5% solution of potassium chrome to this neutral solution and run in silver nitrate solution until the first reddish tinge is observed. Note the reading and repeat the process twice.

Calculation. Suppose 25 cm³ of alkaline sodium chloride solution required a cm³ 0·1M HCl.

Suppose the same neutral solution required b cm³ 0·1M $AgNO_3$.

Concentration of sodium hydroxide

$$= \frac{a}{25} \times 0 \cdot 1M = \frac{a}{25} \times \frac{40}{10} \text{ g/dm}^3$$

Concentration of sodium chloride

$$= \frac{b-a}{25} \times 0 \cdot 1M$$

$$= \frac{b-a}{25} \times \frac{53 \cdot 5}{10} \text{ g/dm}^3$$

Experiment 43. Alternative method for Expt. 42. (Alkaline sodium chloride solution; 0·1M hydrochloric acid; 0·1M silver nitrate; precipitated chalk.)

(*a*) *Alkalinity.* If the concentration of the solution with respect to alkali is required titrate with 0·1M.HCl as indicated above.

(*b*) *Chloride Content.* To a separate 25 cm³ of the solution add a few cm³ of 3M nitric acid (about 3 cm³ should be ample) and then precipitated chalk until some is in excess. If no effervescence is observed on adding a little chalk, then more nitric acid is added. The solution is now neutral and can be titrated with 0·1M $AgNO_3$, using potassium chromate as an indicator. In this case the silver nitrate reading gives the chloride content directly because no hydrochloric acid has been added.

Experiment 44. Estimation of chloride and acid in a solution containing both. (Acidified chloride solution (this solution can be conveniently prepared by putting about 3 g of pure sodium chloride in a 1000 cm³ flask, adding 500 cm³ 0·1M HCl and making up to the mark); 0·1M NaOH; 0·1M $AgNO_3$.)

Silver nitrate cannot be used to titrate an acid solution of a

SILVER NITRATE

chloride using potassium chromate as an indcator. The concentration of the acid is first determined by means of standard alkali and a further quantity of solution is made neutral with calcium carbonate and the chloride content determined from this. Titrate 25 cm^3 of the original solution against sodium hydroxide using phenolphthalein as an indicator. To a further 25 cm^3 add excess of precipitated chalk (see p. 63) and 1 cm^3 of 5% potassium chromate solution as an indicator and titrate with silver nitrate. The concentration of the acid and chloride are calculated from the volumes of alkali and silver nitrate respectively.

Experiment 45. **Estimation of percentage of potassium chlorate in a mixture of potassium chlorate and potassium sulphate.** (Mixture of pure potassium chlorate and potassium sulphate; 0·1M silver nitrate.)

Principle. Potassium chlorate will react with concentrated hydrochloric acid liberating chlorine and is reduced to the chloride which is estimated in the usual way with silver nitrate.

$KClO_3 + 6HCl = KCl + 3H_2O + 3Cl_2$
Molecular weight 122·5.

Method. Weigh out accurately about 3 g of the mixture and place the solid in a large evaporating dish. Add about 20 to 30 cm^3 of concentrated hydrochloric acid, cover the dish with a clock glass and heat gently in a fume chamber for twenty to thirty minutes. There should now be no smell of chlorine. Wash any splashed liquid from the clock glass into the dish, place on a water bath and evaporate carefully down to dryness. Make up the solid left to 250 cm^3 and titrate with the silver nitrate.

Exercises

An asterisk indicates that the particular problem can be attacked without previous preparation of solutions, the concentrations of which have not to be divulged to the student.

(1) [Solutions of sodium carbonate and sodium chloride; 0·1M hydrochloric acid; 0·1M silver nitrate.]
The solution contains only sodium carbonate and sodium chloride. Estimate the separate amounts of each salt in g/dm^3.

(2) [Solution containing sodium chloride, sodium hydroxide and sodium carbonate; 0·1M hydrochloric acid; 0·1M silver nitrate.]
The solution contains sodium chloride, sodium hydroxide and sodium carbonate. Estimate the weight of each compound in 1 dm^3.
(In the estimation of sodium hydroxide the sodium carbonate should first be precipitated by an excess of barium chloride and advantage taken of the fact that the concentration of hydrogen ion required to decompose barium carbonate is greater than that to decolourise phenolphthalein.)

(Cambridge Schol.)

68 THE ESSENTIALS OF VOLUMETRIC ANALYSIS

(3) [Solutions A and B, 0·1M silver nitrate; 0·02M potassium permanganate.]

You are required to determine the percentage proportion of potassium chlorate and potassium chloride in a mixture X of these two compounds.

You are provided with two solutions labelled A and B and with standard solutions of silver nitrate and potassium permanganate. Solution A contains in each 1000 cm³ the residue left on ignition (to constant weight) of 13·57 g of X.

Solution B contains in each 1000 cm³ the product of boiling 13·57 g of X dissolved in sulphuric acid with 60 g $FeSO_4$, the iron (II) sulphate being in excess of that required for the reaction:

$$ClO_3^- + 6Fe^{2+} + 6H^+ = Cl^- + 6Fe^{3+} + 3H_2O$$

(Oxford Schol.)

(4)* To determine the solubility of barium chloride at the temperature of the laboratory. [Barium chloride crystals, 0·1M silver nitrate.]

Crush some crystals of barium chloride and add to water in a boiling tube. Shake vigorously, add more crystals if necessary, and allow to stand with intermittent shakings for about twenty minutes. Work out the weight of solution required to dilute to 250 cm³ in order to be roughly 0·05M. You are told the solubility will be of the order of 400 g/dm³ water at the temperature of the laboratory. Weigh a weighing bottle empty, filter off about the calculated weight into the weighing bottle and weigh again. (If you use a pipette it must be washed out with the saturated solution.) Transfer completely the contents of the weighing bottle to a 250 cm³ flask, empty all washings from the bottle into the flask and make up to 250 cm³. Proceed as in Experiment 41 and calculate the result in g/dm³ solvent.

(5)* To determine the percentage purity of a sample of potassium bromide. [Potassium bromide; 0·1M silver nitrate.]

$$AgNO_3 + KBr = AgBr + KNO_3$$

Molecular weight 119.

Hence tenth molar potassium bromide will contain 11·9 g/dm³. Weigh out accurately about 3 g of potassium bromide. Make up to 250 cm³ with distilled water. Titrate against silver nitrate using potassium chromate solution as an indicator. Calculate the percentage purity of the bromide.

Calculations

(1) 2·5 g of a mixture of potassium chloride and sodium chloride were made up to 250 cm³. 20 cm³ of this solution required 32 cm³ of 0·1M $AgNO_3$ for complete precipitation of the chloride. Calculate the percentage by weight of the two salts in the original mixture.

(2) A solution contained a mixture of sodium carbonate and sodium chloride. 25 cm³ of the solution required 20·4 cm³ of 0·095M HCl to convert the carbonate into the chloride. The same 25 cm³ containing also the chloride converted from the carbonate required 38·7 cm³ 0·1M $AgNO_3$ to precipitate the chloride completely. Calculate the concentration of sodium carbonate and sodium chloride in the original solution.

SILVER NITRATE

(3) A solution of strontium chloride in water was made by dissolving 3·01 g of the crystalline salt in distilled water and the solution was made up to 250 cm³. 20 cm³ of this solution required 17·7 cm³ 0·1M $AgNO_3$ for complete precipitation of the chloride. Calculate the number of molecules of water of crystallization per molecule of $SrCl_2$.

(4) 20 cm³ of a solution of sodium cyanide required 29·8 cm³ of 0·0104M $AgNO_3$ for the formation of the double salt $NaAg(CN)_2$. Calculate the concentration of the sodium cyanide solution in g/dm³.

8. Potassium Thiocyanate

Volhard's Method

BY the use of a solution of either potassium or ammonium thiocyanate the concentration of a solution of silver nitrate can be found. This can be done in acid solution which is sometimes very advantageous (see p. 61). If the solution is neutral it is necessary to add nitric acid to prevent hydrolysis of the iron(III) salt used as indicator. The reaction depends upon the insolubility of silver thiocyanate.

$$Ag^+ + CNS^- = AgSCN \downarrow$$

A solution of an iron(III) salt (usually the alum) is the indicator. If a thiocyanate solution is added to an acidic silver nitrate solution containing iron(III) sulphate the silver thiocyanate is selectively precipitated before the production of any iron(III) thiocyanate because of the very small solubility product of the silver thiocyanate.

As soon as the thiocyanate ions are in excess they react with the iron(III) ions present to produce the blood red coloration.

$$Fe^{3+} + CNS^- = FeSCN^{2+}$$

Volhard's method can be applied to chlorides in acid solution by precipitating all the chloride by the addition of excess silver nitrate solution and determining the excess silver nitrate by titration with potassium thiocyanate.

Note that in this method it is not necessary for the silver nitrate to be in the burette. The thiocyanate solution is run into the silver nitrate solution. The reason for this is that the precipitated silver thiocyanate can adsorb thiocyanate ions, and this would take place to some extent if the potassium or ammonium thiocyanate were in the conical flask and silver nitrate solution were added from the burette.

Experiment 46. Standardisation of potassium thiocyanate solution (Potassium thiocyanate; 0·1M silver nitrate; iron(III) ammonium sulphate.)

Potassium thiocyanate (also ammonium thiocyanate) is a very deliquescent solid and it is impossible to weigh out exactly the amount required for any particular solution.

$$KSCN + AgNO_3 = AgSCN + KNO_3$$

Molecular weight 97.

POTASSIUM THIOCYANATE 71

Hence 1 dm³ of 0·1M solution will contain 9·7 g of anhydrous potassium thiocyanate.

Weigh out about 12 g of the crystals on a watch-glass and transfer the crystals to a measuring flask, add distilled water to make up about 1 dm³ and shake well. Measure out 25 cm³ of $AgNO_3$ (its concentration must be accurately known) into a conical flask, add 2 or 3 cm³ of a 10% solution of iron(III) sulphate and a little dilute nitric acid. The acid prevents the hydrolysis of the iron(III) sulphate, which would otherwise cause the solution to be brownish in colour and so prevent a clear indication of the end-point. Put the thiocyanate solution into a burette and run in the thiocyanate until the first permanent reddish tinge is observed. Repeat the process two or three times.

Calculation. Suppose 25 cm³ of 0·104 M $AgNO_3$ required 23·7 cm³ of KSCN solution.

$$\text{Molarity of KSCN solution} = \frac{25}{23·7} \times 0·104$$
$$= 0·1097$$

Concentration of KSCN solution = 0·1097 × 97 g/dm³
= 10·64 g/dm³

Experiment 47. Estimation of purity of sodium chloride by Volhard's method. (Sodium chloride; 0·1M silver nitrate; 0·05M potassium thiocyanate; iron(III) ammonium sulphate.)

Principle. The sodium chloride is dissolved in water, excess silver nitrate added; the precipitated silver chloride is removed and the excess of silver nitrate is determined by back titration.

Method. Weigh accurately a weighing bottle containing not more than 0·4 g of sodium chloride. Transfer this to a 250 cm³ flask, adding a little distilled water to dissolve the sodium chloride. Reweigh the weighing bottle. Run in 150 cm³ silver nitrate solution, add 2 to 3 cm³ of concentrated nitric acid, replace the stopper and shake until the precipitated silver chloride settles, leaving a clear supernatant liquid. Add distilled water up to the mark and shake thoroughly. Next the solution must be filtered because silver chloride is more soluble than the silver thiocyanate and tends to react with the iron(III) thiocyanate; filter through a dry filter paper. Measure out 50 cm³ of the filtrate by means of a pipette, add 2 or 3 cm³ of the iron(III) solution and run in the potassium thiocyanate solution until the first permanent reddish tinge is observed. Repeat the titration two or three times.

Results. Suppose weight of sodium chloride = 0·376 g
To this 150 cm³ 0·1M $AgNO_3$ were added and the solution was made up to 250 cm³.

72 THE ESSENTIALS OF VOLUMETRIC ANALYSIS

Suppose 50 cm³ of this solution required 34·8 cm³ 0·05M KSCN

$$NaCl + AgNO_3 = AgCl + NaNO_3$$
$$58·5 \text{ g} \quad\quad 10 \text{ dm}^3 \text{ 0·1M}$$

∴ 50 cm³ of the solution ≡ 17·4 cm³ 0·1M KSCN
∴ 250 cm³ of the solution ≡ 87 cm³ 0·1M KSCN
(150 − 87) cm³ of 0·1M AgNO₃ were used up by the chloride,
i.e., 63·0 cm³ 0·1M AgNO₃ ≡ 0·376 g of the impure sodium chloride.
But 10 dm³ 0·1M AgNO₃ ≡ 58·5 g NaCl

∴ 63·0 cm³ 0·1M AgNO₃ ≡ $\frac{58·5}{10\,000} \times 63$ g NaCl
$\quad\quad\quad\quad\quad\quad\quad\quad = 0·3686$ g

Percentage purity $= \frac{0·3686}{0·376} \times 100$
$\quad\quad\quad\quad\quad\quad = 98$

Experiment 48. To determine the percentage of silver in an alloy, e.g., a coin. (Silver alloy; 0·1M potassium thiocyanate.)

For this purpose the alloy should not contain too high a proportion of copper. If too much copper is present the colour of the copper ions interferes. The pre-1920 silver coinage is excellent for this purpose as it contains 92·5% of silver. The 1920–47 coin is not very suitable as it contains 50% Ag; Cu 40%; Ni 5%; Zn 5%. The current silver coinage contains no silver whatsoever.

Weigh accurately a piece of silver of weight about 2 g and introduce the metal carefully into a 250 cm³ beaker. Add about 5 cm³ of distilled water followed by 5 cm³ of concentrated nitric acid. Put the beaker in a fume chamber and allow the metal to dissolve. When the metal has dissolved boil the solution to decompose nitrous acid and then add distilled water, washing the sides of the beaker. Transfer the solution to a measuring flask and add the rinsings from the beaker so that no silver nitrate solution is lost. Make up to the mark and shake well. Measure out 25 cm³ of this solution with a pipette, add 2 or 3 cm³ of the iron(III) solution and titrate with the thiocyanate solution. Note the burette reading at the first permanent reddish tinge, and repeat the titration two or three times.

$$Ag + 2HNO_3 = AgNO_3 + H_2O + NO_2$$
$$AgNO_3 + KSCN = AgSCN + KNO_3$$

Hence $\quad\quad\quad Ag \equiv KSCN$
i.e. $\quad\quad\quad 108 \text{ g} \equiv 10 \text{ dm}^3 \text{ 0·1M}.$

Exercises

An asterisk indicates that the particular problem can be attacked without previous preparation of solutions, the concentrations of which have not to be divulged to the student.

POTASSIUM THIOCYANATE

(1) Solution A contains 5 g of a mixture of potassium nitrate and silver nitrate made up to 250 cm³ of solution. Given 0·1M KSCN determine the concentrations of the two salts in g/dm³.

(2)* Estimation of percentage purity of potassium bromide. [Potassium bromide; 0·1M silver nitrate; 0·05M potassium thiocyanate.]
Repeat Experiment 47, using potassium bromide instead of sodium chloride. Weigh out about 0·7 g of the bromide. As silver bromide is less soluble than silver thiocyanate there is no need to filter off the precipitated silver bromide.

(3)* Estimation of the concentration of mercury(II) nitrate solution.
[Mercury(II) nitrate; 0·1M potassium thiocyanate; sodium nitroprusside.]
Sodium nitroprusside gives a permanent precipitate with mercury(II) nitrate but not with potassium thiocyanate. Make up a solution of mercury nitrate to be about 0·1M. Run this from a burette into 25 cm³ of potassium thiocyanate solution to which a few drops of a 10% solution of sodium nitroprusside have been added. Take as the endpoint the first permanent turbidity.

$$Hg(NO_3)_2 + 2KSCN = Hg(SCN)_2 \downarrow + 2KNO_3$$

(4) [Sodium chloride containing about 1·4 g/250 cm³ solution; arbitrary silver nitrate; unknown ammonium thiocyanate.]
By means of the given pure sodium chloride determine the molarity of the solution of silver nitrate, and use this solution to determine the amount of ammonium thiocyanate (NH_4SCN) in 1 dm³ of solution.

(L.)

Calculations

(1) Determine the concentrations in g/dm³ of potassium cyanide and potassium chloride in a solution containing both from the following data:—
(a) 25 cm³ of the solution required 22·4 cm³ 0·1M $AgNO_3$ for the appearance of the first permanent precipitate. This indicates the end-point of the reaction.
$$AgNO_3 + 2KCN = KAg(CN)_2 + KNO_3$$
(b) To 25 cm³ of the original solution 100 cm³ 0·1M $AgNO_3$ were added and the solution was acidified with nitric acid, filtered and the filtrate required 32·0 cm³ 0·1M KSCN to precipitate the residual silver.

(2) 6 g of pure copper(II) sulphate crystals were dissolved in water and the solution was made up to 250 cm³. 25 cm³ of this solution were treated with a reducing agent and 50 cm³ 0·1M KSCN, and the precipitate of copper(I) thiocyanate was filtered off. The filtrate was equivalent to 25 cm³ of a silver nitrate solution. Calculate the concentration of the silver nitrate solution in g/dm³.

(3) One gramme of an alloy containing silver was dissolved in nitric acid and the solution was made up to 100 cm³. 20 cm³ of this solution required 16·8 cm³ of 0·1M KSCN to precipitate the silver. Calculate the percentage of silver in the alloy.

9. Adsorption Indicators

ADSORPTION indicators play an important part in volumetric analysis: certain dyes can act as indicators for many types of reaction. These adsorption indicators are very sensitive, enabling dilute solutions to be employed. The following example gives an idea of the mechanism of an adsorption indicator.

Charge on Colloidal Particle alters as Reaction is Complete. Many precipitates are colloidal in nature and the colloidal particles can adsorb ions from solution. Ions are not adsorbed by the colloidal particles indiscriminately, for the latter show preference for certain ions over others. The colloidal particles of a silver bromide precipitate will adsorb either bromide or silver ions from solutions containing either of these. Thus silver bromide in potassium bromide solution will adsorb bromide ions but not potassium ions, and silver bromide in silver nitrate solution will adsorb silver ions but not nitrate ions.

Consider the action which takes place when silver nitrate is run into a solution of potassium bromide. So long as the latter is in excess (*i.e.*, before the reaction is complete) there are present potassium ions, bromide ions, nitrate ions and silver bromide particles, *i.e.*, assuming excess KBr

$$2K^+ + 2Br^- + Ag^+ + NO_3^- = AgBr + 2K^+ + NO_3^- + Br^-$$

The colloidal silver bromide adsorbs the bromide ions forming a *negatively* charged complex which we may regard as $[AgBr]Br^-$, *i.e.*, a negatively charged particle. As soon as the reaction is complete there will be no bromide ions in excess and a further drop of silver nitrate will produce a solution containing potassium ions, nitrate ions, *silver ions* and silver bromide particles.

I.e., assuming excess $AgNO_3$

$$K^+ + Br^- + 2Ag^+ + 2NO_3^- = K^+ + 2NO_3^- + Ag^+ + AgBr \downarrow$$

The conditions are now altered. The $[AgBr]Br^-$ complex is decomposed and the silver bromide will now adsorb silver ions to form a complex such as $[AgBr]Ag^+$ which will be *positively* charged. The important point to note is that as the reaction is completed and a further quantity of the reagent (in this case silver nitrate) is added, the colloidal particle changes in sign from negative to positive.

Mechanism of the Indicator. We will consider eosin to have the formula NaEo to indicate that the solution in water contains sodium

ADSORPTION INDICATORS

cations and eosinate anions. These latter impart a pink colour to the solution. (Red ink is a solution of sodium eosinate in water.)

$$NaEo \rightleftharpoons Na^+ + Eo^-$$
$$\text{colourless} \qquad \text{pink}$$

Whilst the bromide adsorbed complex [AgBr]Br$^-$ is present there is no tendency for the colloidal particle to attract to itself the coloured eosinate ions, which therefore *colour the solution but not the precipitate*. When the silver adsorbed complex [AgBr]Ag$^+$ is formed some of the eosinate ions are at once attracted to it by reason of their opposite charge and the precipitate becomes coloured by the eosinate ions. Furthermore, the colour of the eosinate ions when adsorbed by the colloidal particle is not exactly the same as the colour observed when the eosinate ion is in solution (the eosinate ion may combine with the silver ion to form silver eosinate). Hence quite a definite change in colour is observed, and it is the precipitate which becomes coloured. This is an important point in the use of adsorption indicators. It is the precipitate which must be observed and not the solution.

It is also often noticed that the precipitate coagulates as the end-point is reached because this is also the isoelectric point where there is no charge on the colloid particles.

If difficulty is experienced in noticing the end-point, the following method will help. To two conical flasks add some of the reagent, *e.g.*, hydrochloric acid and sufficient calcium carbonate to render the solution neutral, and a few drops of fluorescein. Into flask A run an amount of silver nitrate obviously insufficient to cause any alteration of colour and into flask B sufficient to give a definite change, *i.e.*, an amount greater than that necessary to attain the end-point.

Now place the solution to be titrated by the side of these and run the silver nitrate into it. The first permanent change will be then quite obvious.

The following examples are simple and the indicators used are either eosin (red ink will do) or fluorescein.

Experiment 49. To standardise a solution of hydrochloric acid using fluorescein as indicator. (Approximately 0·1M hydrochloric acid; 0·1M silver nitrate; calcium carbonate.)

Although many adsorption indicators can be used in acid solution fluorescein can only be used in neutral solution.

Measure out from a pipette 25 cm^3 of the hydrochloric acid solution to be standardised and add about 1 g of precipitated calcium carbonate (see Experiment 39 for theoretical consideration) or until there remains a little calcium carbonate undissolved.

Add 2 or 3 drops of fluorescein (which will colour the solution green) and run in the silver nitrate solution from the burette until the

76 THE ESSENTIALS OF VOLUMETRIC ANALYSIS

precipitate (which is yellowish at first) turns pink. Repeat the experiment two or three times. Calculate the molarity of the acid from the equation

$$\underset{\underset{\text{or}}{36\cdot5\text{ g}}}{\text{HCl}} + \text{AgNO}_3 = \text{AgCl} + \text{HNO}_3$$

$$10 \text{ dm}^3 \, 0{\cdot}1\text{M} \quad 10 \text{ dm}^3 \, 0{\cdot}1\text{M}$$

N.B. This equation does not represent the actual reaction which took place. The silver nitrate was titrated against an equivalent solution of calcium chloride.

$$2\text{HCl} + \text{CaCO}_3 = \text{CaCl}_2 + \text{H}_2\text{O} + \text{CO}_2$$
$$\text{CaCl}_2 + 2\text{AgNO}_3 = \text{Ca(NO}_3)_2 + 2\text{AgCl}$$

Experiment 50. To find the molecular weight of potassium bromide.
(0·01M silver nitrate; potassium bromide; eosin.)

Weigh accurately a weighing bottle containing about 0·3 g of pure potassium bromide and transfer this to a 250 cm³ flask and reweigh the bottle. The weighing must be accurate, for a small actual error in weighing this small amount would create a large percentage error. Dissolve the bromide in distilled water, make the solution up to the mark, and shake well. Measure out 25 cm³ of this solution from a pipette, add two or three drops of eosin and titrate with silver nitrate from a burette. Note the reading when the precipitate becomes salmon-pink in colour. Repeat two or three times.

Calculation. Suppose a g of KBr were made up to 250 cm³ of solution and 25 cm³ of this solution required b cm³ of 0·01M AgNO$_3$

b cm³ of 0·01M AgNO$_3 \equiv$ 25 cm³ of KBr solution.

∴ 10 b cm³ 0·01M AgNO$_3 \equiv$ 250 cm³ of KBr solution
$\equiv a$ g of KBr

$$1 \text{ cm}^3 \, 0{\cdot}01\text{M AgNO}_3 \equiv \frac{a}{10b} \text{ g KBr}$$

100 000 cm³ 0·01M AgNO$_3$, *i.e.*, 1 dm³ M

$$\equiv \frac{a}{10b} \times 100\,000 \text{ g KBr}$$

$$= \frac{10\,000 \times a}{b} \text{ g KBr}$$

which is the molecular weight of potassium bromide because the equation for the reaction is

$$\text{KBr} + \text{AgNO}_3 = \text{AgBr} + \text{KNO}_3$$

ADSORPTION INDICATORS

Experiment 51. To estimate the purity of lead nitrate crystals by titration against sodium hydroxide. (0·1M sodium hydroxide; lead nitrate; fluorescein.)

$$Pb(NO_3)_2 + 2NaOH = Pb(OH)_2 + 2NaNO_3$$
331 g   20 dm³ 0·1M

Weigh accurately a weighing bottle containing about 4 g of pure lead nitrate crystals and transfer it to a 250 cm³ flask. Reweigh the weighing bottle. Add distilled water to the crystals, shake, and when dissolved make up to the mark, shake and allow to stand. Fill the burette with the lead nitrate solution. Measure out 25 cm³ of the sodium hydroxide into a conical flask and add two to three drops of a solution of fluorescein. This will colour the solution green. As the lead nitrate solution is run in, the precipitate becomes yellowish in colour. Run in the lead nitrate solution until there is the first permanent pink coloration on the precipitate. Repeat the experiment two or three times.

Experiment 52. To determine the concentration of a potassium thiocyanate solution. (0·1M silver nitrate; approximately 0·1M potassium thiocyanate; fluorescein.)

The solution of potassium thiocyanate should be approximately 0·1M because it is necessary in silver nitrate titrations with adsorption indicators to proceed rapidly. Otherwise the action of light upon the precipitate may obscure the change. Measure out 25 cm³ of the potassium thiocyanate solution, add a few drops of fluorescein and rapidly run in silver nitrate solution until the precipitate coagulates and turns pink. Make two or three more accurate determinations, proceeding more slowly towards the end of the titration.

$$AgNO_3 + KSCN = AgSCN + KNO_3$$
From the equation

$$1 \text{ dm}^3 \ 0{\cdot}1M \ AgNO_3 \equiv 9{\cdot}7 \text{ g KSCN}$$

calculate the concentration of the potassium thiocyanate in g/dm³.

Exercises

An asterisk indicates that the particular problem can be attacked without previous preparations of solutions, the concentrations of which have not to be divulged to the student.

(1)* [Lead acetate; sodium hydroxide; dichlorofluorescein.] Investigate the possibility of using dichlorofluorescein to indicate the end-point of the reaction:—

$$Pb(CH_3CO_2)_2 + 2NaOH = Pb(OH)_2 + 2CH_3CO_2Na$$

Use solutions which are approximately 0·1M and allow the acetate to run into the sodium hydroxide solution.

(2) Concentration of sodium oxalate solution. [Unknown sodium oxalate.]

The lead acetate solution prepared above may be used to estimate the concentration of a solution of sodium oxalate (or any normal oxalate) using fluorescein as adsorption indicator. The precipitate becomes permanently pink in colour at the end-point.

(3) Estimation of bromide and iodide in a mixture. [Mixture of potassium bromide and iodide; di-iodofluorescein, eosin; 0·02M silver nitrate.]

Weigh out accurately about 1 g of the mixture and dissolve in distilled water and make up to 250 cm^3. Titrate 25 cm^3 of the solution.

(a) with silver nitrate using di-iodofluorescein (yellow to pink) which indicates the end-point when the iodide only has been precipitated.

(b) with silver nitrate using eosin which indicates the end-point when the whole of the halide has been precipitated.

Estimate the concentrations of the iodide and bromide in g/dm^3.

ANSWERS TO CALCULATIONS

CHAPTER III. Page 35.

(1) 5·6 g; 9·8 g; 10·6 g.
(2) 40.
(3) 2·8; 0·0285.
(4) 54; 46.
(5) 5·3 g; 8·39 g.

CHAPTER IV. Page 47.

(2) 208 cm^3.
(3) 32·5.
(4) 7.
(5) 99.
(6) 4·99 g; 4·41 g.
(7) 80.
(8) 13·8 g; 19·2 g.

CHAPTER V. Page 53.

(1) 55.
(2) 6·40.

CHAPTER VI. Page 60.

(1) 76.
(2) 8.
(3) 106.
(4) 26·0 cm^3.
(5) 5·06.

CHAPTER VII. Page 68.

(1) 29·8; 70·2.
(2) 4·11 g/dm^3; 4·52 g/dm^3.
(3) 6.
(4) 1·52.

CHAPTER VIII. Page 73.

(1) 11·6 g; 6·91 g.
(2) 17·6.
(3) 90·7.

Table of Atomic Weights 1969.

Based on the Assigned Relative Atomic Mass of $^{12}C = 12$

The values given here apply to elements as they exist in materials of terrestrial origin and to certain artificial elements. When used with due regard to the footnotes they are considered reliable to ± 1 in the last digit, or ± 3 if that digit is in small type.

Name	Symbol	Atomic Number	Atomic Weight	Name	Symbol	Atomic Number	Atomic Weight
Actinium	Ac	89		Mercury	Hg	80	200·59
Aluminium	Al	13	26·9815[a]	Molybdenum	Mo	42	95·94
Americium	Am	95		Neodymium	Nd	60	144·24
Antimony	Sb	51	121·75	Neon	Ne	10	20·179[c]
Argon	Ar	18	39·948[b,c,d,g]	Neptunium	Np	93	237·0482[b,f]
Arsenic	As	33	74·9216[a]	Nickel	Ni	28	58·71
Astatine	At	85		Niobium	Nb	41	92·9064[a]
Barium	Ba	56	137·34	Nitrogen	N	7	14·0067[b,c]
Berkelium	Bk	97		Nobelium	No	102	
Beryllium	Be	4	9·01218[a]	Osmium	Os	76	190·2
Bismuth	Bi	83	208·9806[a]	Oxygen	O	8	15·9994[b,c,d]
Boron	B	5	10·81[c,d,e]	Palladium	Pd	46	106·4
Bromine	Br	35	79·904[c]	Phosphorus	P	15	30·9738[a]
Cadmium	Cd	48	112·40	Platinum	Pt	78	195·09
Caesium	Cs	55	132·9055[a]	Plutonium	Pu	94	
Calcium	Ca	20	40·08	Polonium	Po	84	
Californium	Cf	98		Potassium	K	19	39·102
Carbon	C	6	12·011[b,d]	Praseodymium	Pr	59	140·9077[a]
Cerium	Ce	58	140·12	Promethium	Pm	61	
Chlorine	Cl	17	35·453[c]	Protactinium	Pa	91	231·0359[a,f]
Chromium	Cr	24	51·996[c]	Radium	Ra	88	226·0254[a,f,g]
Cobalt	Co	27	58·9332[a]	Radon	Rn	86	
Copper	Cu	29	63·546[c,d]	Rhenium	Re	75	186·2
Curium	Cm	96		Rhodium	Rh	45	102·9055[a]
Dysprosium	Dy	66	162·50	Rubidium	Rb	37	85·4678[c]
Einsteinium	Es	99		Ruthenium	Ru	44	101·07
Erbium	Er	68	167·26	Samarium	Sm	62	150·4
Europium	Eu	63	151·96	Scandium	Sc	21	44·9559[a]
Fermium	Fm	100		Selenium	Se	34	78·96
Fluorine	F	9	18·9984[a]	Silicon	Si	14	28·086[d]
Francium	Fr	87		Silver	Ag	47	107·868[c]
Gadolinium	Gd	64	157·25	Sodium	Na	11	22·9898[a]
Gallium	Ga	31	69·72	Strontium	Sr	38	87·62[g]
Germanium	Ge	32	72·59	Sulfur	S	16	32·06[d]
Gold	Au	79	196·9665[a]	Tantalum	Ta	73	180·9479[b]
Hafnium	Hf	72	178·49	Technetium	Tc	43	98·9062[f]
Helium	He	2	4·00260[b,c]	Tellurium	Te	52	127·60
Holmium	Ho	67	164·9303[a]	Terbium	Tb	65	158·9254[a]
Hydrogen	H	1	1·0080[b,d]	Thallium	Tl	81	204·37
Indium	In	49	114·82	Thorium	Th	90	232·0381[a,f]
Iodine	I	53	126·9045[a]	Thulium	Tm	69	168·9342[a]
Iridium	Ir	77	192·22	Tin	Sn	50	118·69
Iron	Fe	26	55·847	Titanium	Ti	22	47·90
Krypton	Kr	36	83·80	Tungsten	W	74	183·85
Lanthanum	La	57	138·9055[b]	Uranium	U	92	238·029[b,c,e]
Lawrencium	Lr	103		Vanadium	V	23	50·9414[b,c]
Lead	Pb	82	207·2[d,g]	Wolfram	W	74	183·85
Lithium	Li	3	6·941[c,d,e]	Xenon	Xe	54	131·30
Lutetium	Lu	71	174·97	Ytterbium	Yb	70	173·04
Magnesium	Mg	12	24·305[c]	Yttrium	Y	39	88·9059[a]
Manganese	Mn	25	54·9380[a]	Zinc	Zn	30	65·37
Mendelevium	Md	101		Zirconium	Zr	40	91·22

[a] Mononuclidic element.
[b] Element with one predominant isotope (about 99–100% abundance).
[c] Element for which the atomic weight is based on calibrated measurements.
[d] Element for which variation in isotopic abundance in terrestrial samples limits the precision of the atomic weight given.

Index

A

Accuracy of volumetric analysis 1, 9, 10
Acid, definition of, 11
 see also nitric, oxalic, etc.
Acidimetry & alkalimetry, 21–36
Adsorption indicators, 74–78
Alum, see iron(III) sulphate
Alloy, silver in, 72
Amalgam, zinc, 42, 52
Ammonia, determination of in ammonium salt
 by direct method, 31
 by indirect method, 30
Ammonium oxalate, determination of solubility, 45, 46
Analysis
 of ammonia in ammonium salt
  by direct method, 31
  by indirect method, 30
 of chlorine in bleaching powder, 56, 57
 of hardness in water, 32
 of hydrogen peroxide in solution 43, 44
 of iron in iron wire, 40
 of iron in iron(III) salt, 41, 42, 52
 of iron in iron ore, 52
 of manganese(IV) oxide in pyrolusite, 46
 of potassium chlorate in sample, 51
 of potassium thiocyanate in solution, 77
 of silver in alloy, 72
 of sodium sulphite in sample, 58, 59
 see also mixture, purity, water of crystallization, molecular weight determination
Answers to calculations, 79
Avogadro constant, 2
Azolitmin, 20

B

Back titration, 26
Barium chloride, water of crystallization determination, 64, 65
Base, definition of, 11
Bleaching powder, chlorine in, 6, 575
Borax, 23, 24
Burette, 6–8

C

Calcium carbonate
 molecular weight determination, 26
 used to standardise nitric acid, 25
Calculations
 acidimetry & alkalimetry, 35, 36
 iodine & sodium thiosulphate, 60
 potassium dichromate, 53
 potassium permanganate, 47, 48
 potassium thiocyanate, 73
 silver nitrate, 68, 69
Caustic soda, see sodium hydroxide
Chlorine in bleaching powder, 56, 57
Copper sulphate, purity of, 57, 58

D

Definition,
 of acid, 11
 of base, 11
 of molar solution, 1
 of mole, 2
 of neutralisation, 12
 of standard solution, 1
Diphenylamine indicator, 49–53
Direct method of analysis, 31
Dissociation constant, 12
Double indicator method of analysis, 27–29

E

Electrolytes, 12
End point,
 colour change of precipitate as, 74–77
 inaccuracy of, 9
 of potassium permanganate titration, 37
 precipitation as, 61

INDEX

Eosin, 74, 76
Exercises,
  acidimetry & alkalimetry, 32–35
  adsorption indicators, 77, 78
  iodine & sodium thiosulphate, 59
  potassium dichromate, 52, 53
  potassium permanganate, 47
  potassium thiocyanate, 72, 73
  silver nitrate, 67, 68

F

Fluorescein, 75, 77

H

Hardness, temporary, in water, 32
Hydrochloric acid, standardised,
  with borax, 23, 24
  with silver nitrate, 62, 63
  and fluorescein, 75, 76
  with sodium carbonate, 21–23
Hydrogen peroxide, estimation of concentration, 43, 44
Hydrolysis, 13

I

Icelandic spar, 25
Impurity, causing error in analysis, 9
Inaccuracy of volumetric analysis, 8, 9
Indicators,
  for acid–alkali titrations, 11–20
    choice of 17–20
    double method, 27
    theory of, 16
    see also litmus, methyl orange etc.
  adsorption, 74–78
    see also eosin, fluorescein
  diphenylamine, 49–53
  iron(III) salt, 70–73
  potassium chromate, 61–67
  potassium permanganate as own, 37–48
  silver nitrate as own, 61
  starch, 54–59
Indirect method of analysis, 30
Instruments of volumetric analysis, 3
  inaccuracy of, 8, 9
Iodine,
  reaction with sodium thiosulphate, 54–60

Iodine—cont.
  standard solution of, 56
Ionisation, 11
  of water, 13
Iron, estimation of,
  in iron ore, 52
  in iron(III) salt, 41, 42, 52
  in iron wire, 40
Iron(II) ammonium sulphate,
  used to standardise potassium permanganate, 38
  water of crystallization determination, 45
Iron ore, estimation of iron in, 52
Iron(II) sulphate, determination of water of crystallization, 40
Iron(III) sulphate,
  as indicator, 70–73
  iron in, 41, 42, 52

L

Lead nitrate, purity of, 77
Litmus, 19, 20

M

Measuring flask, 4–6
Methyl orange,
  indicator in analytical experiments, 22–23, 26–32
  modified with xylene cyanol FF, 20
  pH range, 17, 19
  preparation of, 20
Methyl red,
  pH range, 17, 19
  preparation of, 20
Mixture, analysis of,
  chloride & acid in, 66, 67
  chloride & alkali in, 65, 66
  oxalic acid & oxalate in, 43
  potassium chlorate in, 67
  potassium & sodium chlorides, 63–64
  potassium sulphate & potassium permanganate in, 45
  sodium carbonate & sodium hydroxide in,
    by double indicator method, 27–29
    by Winkler method, 29, 30
Molar solution,
  definition of, 1
  water of crystallization and concentration of, 2

INDEX

Mole, definition of, 2
Molecular weight determination,
  of calcium carbonate, 26
  of potassium bromide, 76

N

Neutralisation, 12
Nitric acid, standardisation of, 25, 26

O

Oxalic acid,
  analysis of in mixture, 43
  used to standardise sodium hydroxide, 24

P

pH value, 14–16
  range & indicators, 17
Phenolphthalein,
  indicator in analytical experiments, 24, 25, 28, 29
  pH range & use, 18
  preparation of, 20
Pipette, 6
Potassium bromide, molecular weight determination, 76
Potassium chlorate,
  estimation in sample, 51
  in mixture, 67
Potassium chloride estimation in mixture, 63, 64
Potassium chromate,
  as indicator, 61–67
  purity of, 51
Potassium dichromate, 49–53
Potassium hydrogen oxalate, 46
Potassium iodate, used to standardise sodium thiosulphate, 55
Potassium permanganate, 37–48
  estimation of in mixture, 45
  standardisation with iron(II) salt, 38
  with sodium oxalate, 39, 40
  used to standardise sodium thiosulphate, 55
Potassium sulphate, estimation of in mixture, 45
Potassium thiocyanate,
  as analytical reagent, 70–73
  estimation of concentration, 77
  standardisation of 70, 71

Precipitation as end point in titration, 61, 74–77
Purity,
  of copper sulphate, 57, 58
  of lead nitrate, 77
  of potassium chlorate, 51
  of potassium chromate, 51
  sodium chloride by Volhard's method, 71, 72
  of sodium nitrite, 44, 45
  of sodium sulphite, 58, 59
  of tin, 50, 51
Pyrolusite, estimation of manganese(IV) oxide in, 46

Q

Questions,
  on acidimetry & alkalimetry, 35
  on iodine & sodium thiosulphate, 60
  potassium dichromate, 53
  on potassium permanganate, 47
  on potassium thiocyanate, 73
  on silver nitrate, 68

S

Silver in alloy, 72
Silver nitrate,
  in estimation of chloride, 61–69, 71, 72
  with potassium thiocyanate, 70–72
  standardisation of, 62
Sodium carbonate,
  estimation in mixture, 27–29
  determination of water of crystallization, 27
  used to standardise hydrochloric acid, 21–23
Sodium chloride,
  estimation of in mixture of chlorides, 63, 64
  in mixture with acid, 66, 67
  in mixture with alkali, 65, 66
  purity determined by Volhard's method, 71, 72
Sodium hydroxide,
  estimation of in mixture, 27, 29
  standardised with oxalic acid, 24
  with succinic acid, 24, 25
Sodium metaborate, used to standardise hydrochloric acid, 23, 24
Sodium nitrite, purity of, 44, 45

INDEX

Sodium oxalate, used to standardise potassium permanganate, 39, 40
Sodium sulphite, estimation in sample, 58, 59
Sodium thiosulphate & iodine, 54–60
   standardised with potassium iodate, 55
   with potassium permanganate, 55
Solubility of ammonium oxalate, 45, 46
Spathic iron ore, iron in, 52
Standard solution,
   characteristics of good, 21
   definition, 1
   of iodine, 56
   of potassium dichromate, 49
   of potassium permanganate, 38
   of silver nitrate, 62
   of sodium thiosulphate, 54
Standardisation,
   of hydrochloric acid 21, 23, 24, 62, 63, 75, 76
   of nitric acid, 25, 26
   of potassium permanganate, 38–40
   of potassium thiocyanate, 70, 71
   of silver nitrate, 62
   of sodium hydroxide, 24, 25
   of sodium thiosulphate, 55
Starch,
   as indicator, 54–59
   preparation of, 54
Succinic acid, used to standardise sodium hydroxide 24

T

Tin, purity of, 50, 51

V

Volhard's method, 70–72
Volumetric analysis,
   accuracy of, 9, 10
   inaccuracy of, 8, 9
   instruments of, 3
   see also burette, pipette etc.

W

Washing soda, see sodium carbonate
Water,
   ionisation of, 13
   temporary hardness, determination of, 32
Water of crystallization,
   determination of in barium chloride, 64, 65
   in iron(II) ammonium sulphate, 45
   in iron(II) sulphate, 40
   in washing soda, 27
   and molar solution concentration, 2
Weighing bottle, 3, 4
Winkler method of analysis, 29, 30

X

Xylene cyanol FF, modification of methyl orange, 20

Z

Zinc amalgam, 42, 52

LOGARITHMS 87

	0	1	2	3	4	5	6	7	8	9	1	2	3	4	5	6	7	8	9
10	0000	0043	0086	0128	0170	0212	0253	0294	0334	0374	4	8	12	17	21	25	29	33	37
11	0414	0453	0492	0531	0569	0607	0645	0682	0719	0755	4	8	11	15	19	23	26	30	34
12	0792	0828	0864	0899	0934	0969	1004	1038	1072	1106	3	7	10	14	17	21	24	28	31
13	1139	1173	1206	1239	1271	1303	1335	1367	1399	1430	3	6	10	13	16	19	23	26	29
14	1461	1492	1523	1553	1584	1614	1644	1673	1703	1732	3	6	9	12	15	18	21	24	27
15	1761	1790	1818	1847	1875	1903	1931	1959	1987	2014	3	6	8	11	14	17	20	22	25
16	2041	2068	2095	2122	2148	2175	2201	2227	2253	2279	3	5	8	11	13	16	18	21	24
17	2304	2330	2355	2380	2405	2430	2455	2480	2504	2529	2	5	7	10	12	15	17	20	22
18	2553	2577	2601	2625	2648	2672	2695	2718	2742	2765	2	5	7	9	12	14	16	19	21
19	2788	2810	2833	2856	2878	2900	2923	2945	2967	2989	2	4	7	9	11	13	16	18	20
20	3010	3032	3054	3075	3096	3118	3139	3160	3181	3201	2	4	6	8	11	13	15	17	19
21	3222	3243	3263	3284	3304	3324	3345	3365	3385	3404	2	4	6	8	10	12	14	16	18
22	3424	3444	3464	3483	3502	3522	3541	3560	3579	3598	2	4	6	8	10	12	14	15	17
23	3617	3636	3655	3674	3692	3711	3729	3747	3766	3784	2	4	6	7	9	11	13	15	17
24	3802	3820	3838	3856	3874	3892	3909	3927	3945	3962	2	4	5	7	9	11	12	14	16
25	3979	3997	4014	4031	4048	4065	4082	4099	4116	4133	2	3	5	7	9	10	12	14	15
26	4150	4166	4183	4200	4216	4232	4249	4265	4281	4298	2	3	5	7	8	10	11	13	15
27	4314	4330	4346	4362	4378	4393	4409	4425	4440	4456	2	3	5	6	8	9	11	13	14
28	4472	4487	4502	4518	4533	4548	4564	4579	4594	4609	2	3	5	6	8	9	11	12	14
29	4624	4639	4654	4669	4683	4698	4713	4728	4742	4757	1	3	4	6	7	9	10	12	13
30	4771	4786	4800	4814	4829	4843	4857	4871	4886	4900	1	3	4	6	7	9	10	11	13
31	4914	4928	4942	4955	4969	4983	4997	5011	5024	5038	1	3	4	6	7	8	10	11	12
32	5051	5065	5079	5092	5105	5119	5132	5145	5159	5172	1	3	4	5	7	8	9	11	12
33	5185	5198	5211	5224	5237	5250	5263	5276	5289	5302	1	3	4	5	6	8	9	10	12
34	5315	5328	5340	5353	5366	5378	5391	5403	5416	5428	1	3	4	5	6	8	9	10	11
35	5441	5453	5465	5478	5490	5502	5514	5527	5539	5551	1	2	4	5	6	7	9	10	11
36	5563	5575	5587	5599	5611	5623	5635	5647	5658	5670	1	2	4	5	6	7	8	10	11
37	5682	5694	5705	5717	5729	5740	5752	5763	5775	5786	1	2	3	5	6	7	8	9	10
38	5798	5809	5821	5832	5843	5855	5866	5877	5888	5899	1	2	3	5	6	7	8	9	10
39	5911	5922	5933	5944	5955	5966	5977	5988	5999	6010	1	2	3	4	5	7	8	9	10
40	6021	6031	6042	6053	6064	6075	6085	6096	6107	6117	1	2	3	4	5	6	8	9	10
41	6128	6138	6149	6160	6170	6180	6191	6201	6212	6222	1	2	3	4	5	6	7	8	9
42	6232	6243	6253	6263	6274	6284	6294	6304	6314	6325	1	2	3	4	5	6	7	8	9
43	6335	6345	6355	6365	6375	6385	6395	6405	6415	6425	1	2	3	4	5	6	7	8	9
44	6435	6444	6454	6464	6474	6484	6493	6503	6513	6522	1	2	3	4	5	6	7	8	9
45	6532	6542	6551	6561	6571	6580	6590	6599	6609	6618	1	2	3	4	5	6	7	8	9
46	6628	6637	6646	6656	6665	6675	6684	6693	6702	6712	1	2	3	4	5	6	7	7	8
47	6721	6730	6739	6749	6758	6767	6776	6785	6794	6803	1	2	3	4	5	5	6	7	8
48	6812	6821	6830	6839	6848	6857	6866	6875	6884	6893	1	2	3	4	4	5	6	7	8
49	6902	6911	6920	6928	6937	6946	6955	6964	6972	6981	1	2	3	4	4	5	6	7	8
50	6990	6998	7007	7016	7024	7033	7042	7050	7059	7067	1	2	3	3	4	5	6	7	8
51	7076	7084	7093	7101	7110	7118	7126	7135	7143	7152	1	2	3	3	4	5	6	7	8
52	7160	7168	7177	7185	7193	7202	7210	7218	7226	7235	1	2	2	3	4	5	6	7	7
53	7243	7251	7259	7267	7275	7284	7292	7300	7308	7316	1	2	2	3	4	5	6	6	7
54	7324	7332	7340	7348	7356	7364	7372	7380	7388	7396	1	2	2	3	4	5	6	6	7
	0	1	2	3	4	5	6	7	8	9	1	2	3	4	5	6	7	8	9

LOGARITHMS

	0	1	2	3	4	5	6	7	8	9	1	2	3	4	5	6	7	8	9
55	7404	7412	7419	7427	7435	7443	7451	7459	7466	7474	1	2	2	3	4	5	5	6	7
56	7482	7490	7497	7505	7513	7520	7528	7536	7543	7551	1	2	2	3	4	5	5	6	7
57	7559	7566	7574	7582	7589	7597	7604	7612	7619	7627	1	2	2	3	4	5	5	6	7
58	7634	7642	7649	7657	7664	7672	7679	7686	7694	7701	1	1	2	3	4	4	5	6	7
59	7709	7716	7723	7731	7738	7745	7752	7760	7767	7774	1	1	2	3	4	4	5	6	7
60	7782	7789	7796	7803	7810	7818	7825	7832	7839	7846	1	1	2	3	4	4	5	6	6
61	7853	7860	7868	7875	7882	7889	7896	7903	7910	7917	1	1	2	3	4	4	5	6	6
62	7924	7931	7938	7945	7952	7959	7966	7973	7980	7987	1	1	2	3	3	4	5	6	6
63	7993	8000	8007	8014	8021	8028	8035	8041	8048	8055	1	1	2	3	3	4	5	5	6
64	8062	8069	8075	8082	8089	8096	8102	8109	8116	8122	1	1	2	3	3	4	5	5	6
65	8129	8136	8142	8149	8156	8162	8169	8176	8182	8189	1	1	2	3	3	4	5	5	6
66	8195	8202	8209	8215	8222	8228	8235	8241	8248	8254	1	1	2	3	3	4	5	5	6
67	8261	8267	8274	8280	8287	8293	8299	8306	8312	8319	1	1	2	3	3	4	5	5	6
68	8325	8331	8338	8344	8351	8357	8363	8370	8376	8382	1	1	2	3	3	4	4	5	6
69	8388	8395	8401	8407	8414	8420	8426	8432	8439	8445	1	1	2	2	3	4	4	5	6
70	8451	8457	8463	8470	8476	8482	8488	8494	8500	8506	1	1	2	2	3	4	4	5	6
71	8513	8519	8525	8531	8537	8543	8549	8555	8561	8567	1	1	2	2	3	4	4	5	5
72	8573	8579	8585	8591	8597	8603	8609	8615	8621	8627	1	1	2	2	3	4	4	5	5
73	8633	8639	8645	8651	8657	8663	8669	8675	8681	8686	1	1	2	2	3	4	4	5	5
74	8692	8698	8704	8710	8716	8722	8727	8733	8739	8745	1	1	2	2	3	4	4	5	5
75	8751	8756	8762	8768	8774	8779	8785	8791	8797	8802	1	1	2	2	3	3	4	5	5
76	8808	8814	8820	8825	8831	8837	8842	8848	8854	8859	1	1	2	2	3	3	4	5	5
77	8865	8871	8876	8882	8887	8893	8899	8904	8910	8915	1	1	2	2	3	3	4	4	5
78	8921	8927	8932	8938	8943	8949	8954	8960	8965	8971	1	1	2	2	3	3	4	4	5
79	8976	8982	8987	8993	8998	9004	9009	9015	9020	9025	1	1	2	2	3	3	4	4	5
80	9031	9036	9042	9047	9053	9058	9063	9069	9074	9079	1	1	2	2	3	3	4	4	5
81	9085	9090	9096	9101	9106	9112	9117	9122	9128	9133	1	1	2	2	3	3	4	4	5
82	9138	9143	9149	9154	9159	9165	9170	9175	9180	9186	1	1	2	2	3	3	4	4	5
83	9191	9196	9201	9206	9212	9217	9222	9227	9232	9238	1	1	2	2	3	3	4	4	5
84	9243	9248	9253	9258	9263	9269	9274	9279	9284	9289	1	1	2	2	3	3	4	4	5
85	9294	9299	9304	9309	9315	9320	9325	9330	9335	9340	1	1	2	2	3	3	4	4	5
86	9345	9350	9355	9360	9365	9370	9375	9380	9385	9390	1	1	2	2	3	3	4	4	5
87	9395	9400	9405	9410	9415	9420	9425	9430	9435	9440	0	1	1	2	2	3	3	4	4
88	9445	9450	9455	9460	9465	9469	9474	9479	9484	9489	0	1	1	2	2	3	3	4	4
89	9494	9499	9504	9509	9513	9518	9523	9528	9533	9538	0	1	1	2	2	3	3	4	4
90	9542	9547	9552	9557	9562	9566	9571	9576	9581	9586	0	1	1	2	2	3	3	4	4
91	9590	9595	9600	9605	9609	9614	9619	9624	9628	9633	0	1	1	2	2	3	3	4	4
92	9638	9643	9647	9652	9657	9661	9666	9671	9675	9680	0	1	1	2	2	3	3	4	4
93	9685	9689	9694	9699	9703	9708	9713	9717	9722	9727	0	1	1	2	2	3	3	4	4
94	9731	9736	9741	9745	9750	9754	9759	9763	9768	9773	0	1	1	2	2	3	3	4	4
95	9777	9782	9786	9791	9795	9800	9805	9809	9814	9818	0	1	1	2	2	3	3	4	4
96	9823	9827	9832	9836	9841	9845	9850	9854	9859	9863	0	1	1	2	2	3	3	4	4
97	9868	9872	9877	9881	9886	9890	9894	9899	9903	9908	0	1	1	2	2	3	3	4	4
98	9912	9917	9921	9926	9930	9934	9939	9943	9948	9952	0	1	1	2	2	3	3	4	4
99	9956	9961	9965	9969	9974	9978	9983	9987	9991	9996	0	1	1	2	2	3	3	3	4
	0	1	2	3	4	5	6	7	8	9	1	2	3	4	5	6	7	8	9